Strategies for Research in Healthcare Settings: Challenges and Opportunities

Sub-Proceedings of EDRA53

Greenville, South Carolina, USA
1-4 June 2022

ISBN: 978-1-7138-6843-9

Printed from e-media with permission by:

Curran Associates, Inc.
57 Morehouse Lane
Red Hook, NY 12571

Some format issues inherent in the e-media version may also appear in this print version.

The Environmental Design Research Association (EDRA)

2010 Pear St.
#1060 Pinole, CA
94564-9998

www.edra.org

Additional copies of this publication are available from:

Curran Associates, Inc.
57 Morehouse Lane
Red Hook, NY 12571 USA
Phone: 845-758-0400
Fax: 845-758-2633
Email: curran@proceedings.com
Web: www.proceedings.com

Strategies for Research in Healthcare Settings: Challenges and Opportunities

EDRA53 Local Committee

Anjali Joseph, Ph.D. -- *Co-Chair*
Dina Battisto, Ph.D. -- *Co-Chair*
B.D. Wortham-Galvin, Ph.D. -- *Co-Chair*
James Lyndon McCracken -- *Support*

EDRA 53 Program Committee

Lindsay McCunn, Ph.D. -- *Chair*
Alexandre H. Champagne
David de la Peña, Ph.D.
Deni Ruggeri, Ph.D.
Hui Cai, Ph.D.
Rich Wener, Ph.D.
Sharmin Kader, Ph.D.

PO Box 43023, Washington, DC 20010 - USA

www.edra.org

Editors

Ann Sloan Devlin, PhD
Sharmin Kader, PhD

Authors

Ann Sloan Devlin
Rebecca Davis
Margaret Calkins
Kara Freihoefer
Renae Rich
Saif Haq
Matthew H. E. M. Browning
Fu Li
Nicholas Watkins
April Spivack
Kyle Sellers
Anita Grabowska
Lorissa MacAllister
Ellen Taylor

Sponsor

Gensler

Clemson Student Team

Monica Gripko
Sara Kennedy
Hadley Taylor
Taylor Wahler
Isabel Korn
Swati Goel
Mina Ardekani
Devi Soman

Special Thanks:

A special thanks to Anjali Joseph, without whose encouragement the intensive would not have materialized and whose students facilitated the graphic design of the document and to Nick Watkins of Gensler for taking the initiative to sponsor the publication.

TABLE OF CONTENTS

EDITORS

Ann Sloan Devlin is the May Buckley Sadowski '19 Professor Emerita of Psychology at Connecticut College. She received her B.A., M.A., and Ph.D. from the University of Michigan and joined Connecticut College in 1973, retiring in 2022 after 49 years of teaching undergraduate students. A former Woodrow Wilson Fellow, Devlin is the recipient of several Mellon Foundation grants and with colleagues from Portugal received an Academy of Architecture for Health Foundation grant.

Her expertise lies in environmental psychology, particularly in the creation of more humanistic environments for healthcare, including psychotherapy offices and facilities that house the elderly. She also specializes in way-finding and frequently writes about issues in evidence-based design. A prolific writer, Devlin is the author of seven books, including *The Research Experience: Planning, Conducting and Reporting Research, Transforming the Doctor's Office: Principles from Evidence-based design, and What Americans Build and Why: Psychological Perspectives.* Her articles, many of which include student co-authors, have been published in numerous journals, including *Environment and Behavior, Health Environments Research & Design Journal, The Journal of Counseling Psychology, The Journal of Environmental Psychology*, and the *Journal of Applied Social Psychology*. She is a former editor-in-chief of the journal *Environment and Behavior* and was elected to the Board of Directors of the Environmental Design Research Association in 2000. In 2020 she received the Environmental Design Research Association Career Award.

Sharmin Kader is an architectural designer, researcher, and mentor. She received her doctoral degree from the University of Kansas, and her master's degree from Texas A&M University focusing on healthcare environments. Currently, she is working as a research scholar in the Healthcare Design Program at Kent State University. She worked in multiple design firms including TreanorHL as a lead design researcher. Sharmin's project areas mostly focus on healthcare, higher education, and housing. She has refined her skills in post- occupancy evaluation (POE) and developed the first-ever POE tool for the hospice environment. Sharmin is the recipient of many prestigious awards and honors, such as the AIA Arthur N. Tuttle Jr. Graduate Fellowship in Health Facility Planning & Design, and the Outstanding International Woman Student Award from the University of Kansas. Three of her professional projects received the Certificate of Research Excellence from EDRA. She served on the Board of the Environment Design Research Association (EDRA) as a Chair for two consecutive terms.

INTRODUCTION

EDRA 53 "Health In All Design: Promoting Health, Equity, Sustainability, and Resilience Through Environmental Design" provided a forum to examine how the design of environments contributes to health. Pedagogies & Methods, one of the six conference themes, in turn gave us the opportunity to examine the myriad ways in which research evidence contributes to improved health. Support from Gensler, the global architecture, design and planning firm, enabled us to create this sub-proceedings focused specifically on different research approaches to the design of healthcare environments. The future of healthcare is about change, and a focus on experience, engagement, and outcomes is creating a context that is ripe for innovation. From virtual health platforms and retail clinics, to community health centers and academic medicine, Gensler shapes healthy experiences and outcomes. Leveraging the power of research-informed and purposeful user-centered design, the firm is unlocking design solutions and strategies that are defining the next chapter in the healthcare industry.

VISION FOR THE INTENSIVE

Encouraged by conference co-chair Anjali Joseph, we developed a half-day intensive "Strategies for Research in Healthcare Settings: Challenges and Opportunities" to focus on the variety of approaches that are used in research on healthcare environments. Research in healthcare settings is challenging for a variety of reasons, including hospital bureaucracy, institutional review boards (IRBs), and numerous other gatekeepers. As a result, researchers are often forced to make compromises that undermine the rigor of their methodology, in the process adopting approaches with questionable internal validity. In addition, it is not clear that all practitioners have received sufficient training in research methods to recognize what those threats to internal validity might be (e.g., confounding variables). Thus, studies often move forward that make only a limited contribution to the research because of the drawbacks in their research design.

The vision of the intensive was to provide information to enable conference attendees to improve the methodological rigor of their research by discussing the kinds of barriers the presenters have encountered and the solutions they have developed to address these challenges. This compendium of papers from the intensive provides the opportunity to share this information with an audience beyond EDRA 53 attendees.

Given that the research question should determine the research strategy, the intensive focused on the kinds of questions suitable to answer with a given strategy. One of the major contributions of the intensive (and this compendium) is that a continuum of data collection strategies and research designs is presented, from experimental lab studies and randomized controlled trials to occupancy evaluation designs, national survey research, and the contributions of secondary research. Beyond the types of research questions addressed, presenters include discussions of sample selection, specific analytic strategies such as space syntax and approaches such as virtual reality. The eight papers contained here provide a current snapshot of the kinds of research approaches that are being used to advance healthcare and should be beneficial to a variety of readers, from students to practitioners. The order of the eight papers reflects a continuum starting with tight experimental control (the laboratory study) and ending with secondary research.

EXECUTIVE SUMMARY OF PAPERS

1. Ann Sloan Devlin

Ann Sloan Devlin *is the May Buckley Sadowski '19 professor emerita of psychology at Connecticut College and EDRA 2020 Career Award recipient. Her current areas of research include the role of architecture in health-care outcomes, perceived control as a mediator of stress in health-care settings, and perceptions of therapists' offices.*

The first paper in this compendium examines how the small laboratory experiment can be used as a platform on which to base larger field experiments and studies. Measured on any dimension, field research in healthcare environments is expensive. Before moving into the field, small, tightly controlled experimental research can provide evidence about which healthcare design interventions may be advantageous for healthcare outcomes, setting the stage to test the interventions in situ. Drawing on a body of research dealing with inpatient and outpatient settings, this paper by Ann S. Devlin emphasizes the benefits and drawbacks in three areas: research design, approaches to use of visual stimuli, and participants in online studies. In 1966, Rachel Kaplan, a methodological pioneer, published an article in the EDRA Proceedings focusing on what can be learned from small experiments. This paper returns to the small experiment to focus on the benefits (and challenges) of doing laboratory work with implications for real world applications in healthcare.

2. Rebecca Davis & Margaret Calkins

Rebecca Davis *is a Professor at Grand Valley State University, Kirkhof College of Nursing in Grand Rapids, MI. Through grants from the National Institutes on Aging and the John A. Hartford Foundation, she conducts research on purposeful environmental design to improve wayfinding and overall functioning for older adults with cognitive impairment.*

Dr. Calkins *is internationally recognized as a creative, dynamic leader, trainer and researcher in the field of environments for elders. She is Board Chair of IDEAS Institute, an independent research institute dedicated to exploring the therapeutic potential of the environment--social and organizational as well as physical--particularly as it relates to frail and impaired older adults.*

In the second paper, Rebecca Davis and Margaret Calkins tackle the challenges of measuring the impact of environmental features on wayfinding. Specifically, the work of their Wayfinding Study Team focuses on methods to develop a wayfinding intervention for older adults in long-term care settings. With an emphasis on the importance of environmental cues, specifically signage and landmarks, they explain their approach to testing methods first in the laboratory through virtual reality and then extending their work to the field. The paper discusses the advantages and disadvantages of such VR testing and the design of their current work evaluating the interventions in the field in a randomized controlled study. The paper further explains the challenges of measuring environmental complexity, a critical feature in wayfinding.

3. Kara Freihoefer & Renae Rich

***Kara Freihoefer**, PhD, is the Director of Research at HGA Architects and Engineers where she focuses on building a research-driven practice to better inform and educate clients and colleagues on the human impact of design decisions. In 2021, Kara was recognized as Healthcare Design magazine's researcher of the year as part of the HCD10 class.*

***Renae Rich**, as the Occupant Evaluation Manager at HGA, oversees a program within the Design Insight Group to build HGA's reputation for research and measured results through the systematic evaluation of design projects in terms of occupant experience and satisfaction across all our market sectors.*

In the third paper, Freihoefer and Rich from HGA Architects and Engineers argue that firms and their clients will benefit by prioritizing occupancy evaluations as a strategic initiative. The emphasis on occupancy evaluation has the potential to differentiate the firm as a "curious, research-driven organization," reflecting its use of data, evidence, and ultimately original insights to inform design. The paper looks at the history of occupancy evaluation (formerly post-occupancy evaluation) and the value of such research strategies to improving design. The authors highlight the numerous challenges to incorporating occupancy evaluation as a component of design, given that such research is not viewed as part of the normal design process. The authors present an argument that occupancy evaluation can range from the simple (e.g., building walk-through) to the multi-faceted (e.g., triangulated multi-method approach) and include a range of data gathering approaches (subjective, objective, retrospective, and prospective). The paper demonstrates how to make clear the value of occupancy evaluation to both firms and clients.

4. Saif Haq

***Dr. Saif Haq** is an award-winning researcher, educator, administrator, and design professional. He is the founder-director of the Design for Health program at Texas Tech University and a board member of 'Architecture Research Centers Consortium'. At EDRA he is the co-founder of EDRAMOVE.*

One of the techniques mentioned in a number of papers in the compendium is space syntax, a technique that allows researchers to measure environmental properties arising from the connections of a given space to every other space in the same layout. Saif Haq's paper explains space syntax in an understandable way and identifies the five fundamental decisions a researcher must make in a methodological approach to space syntax. Starting with the theoretical concepts that underly space syntax, he moves to how space syntax is used as a methodology of measurement. The paper uses numerous figures to illustrate the principles being discussed. Concepts include unit spaces, axial lines, the spatial system, accessibility, visibility, closeness, and betweenness. The paper explains how space syntax captures a synchronous model of movement through space, which is an asynchronous experience.

5. Matthew Browning & Fu Li

***Matthew Browning** is the Virtual Reality & Nature Lab Director and Associate Professor in the Department of Parks, Recreation and Tourism Management at Clemson University. His research examines the effects of physical and simulated environments on human health. His Lab's mission is to strengthen the frequency and richness of human-nature interactions.*

***Fu Li** is a Ph.D. student in the Department of Parks, Recreation and Tourism Management at Clemson University. His research interests include topics in Environmental Perception, Behavior Studies, and Landscape & Urban Planning. He is the former Senior Research Associate in the College of Architecture and Landscape Architecture, Peking University.*

The paper by Browning and Li presents strategies for studying human health through the use of virtual nature. Including a summary of the systematic reviews available on VR as they relate to restorative environments, the paper reviews what is now known about the impact on human users of exposure to nature imagery and the kinds of dependent variables that are used (e.g., subjective assessments such as mood; objective indices such as blood pressure or alpha waves. The authors discuss the sources of content for VR studies (e.g., professional photographs, drone capture, gaming engines) and examine the gaps in the literature and important directions for future research, including five methodological implications (e.g., developing better recruitment methods).

6. Nicholas Watkins, April Spivack, Kyle Sellers & Anita Grabowska

*As Gensler's Health Sector Research Lead, **Nicholas Watkins**, PhD, seeks out hidden dimensions of humane settings. The work focuses on the interactions between humans and their built environments that reflect excellence in design and contribute to health wellness and well-being. He has served as an EDRA board member and Chair.*

***April J. Spivack** is an Associate Professor of Entrepreneurship, Management, & Organisation at Hanken School of Economics. Her research uses quantitative and qualitative methods to examine first-person experiences, the interplay of context and person, and creativity and well-being outcomes for entrepreneurs and other organizational members.*

***Kyle Sellers** works in design research, exploring an array of issues related to urban planning, commercial real estate, workplace dynamics, and healthcare. He holds a Master's Degree from the Johns Hopkins University, and he is currently pursuing a Master's of Science in Population Health Informatics from the City University of New York.*

***Anita Grabowska** is a research strategist for Gensler, who specializes in people-centric research to understand the human experience, with expertise in quantitative and qualitative methodologies. She holds a Master's Degree from the University of Virginia.*

The paper by Watkins, Spivack, Sellers, and Grabowska from Gensler moves us from examples that deal with relatively small samples to a discussion of a nationwide study of a US sample of over 3500 outpatients. Because the authors argue that the choice of the panel provider to obtain participants is the most important step in the research process, they examine the criteria for that selection in depth. A useful aspect of this examination is to evaluate such possibilities as Amazon Mechanical Turk and Kantar Profiles. A second focus of the paper is the issue of questionnaire development and administration and how the quality of data submitted by participants can be improved, especially given the problem of professional survey takers who may not be attentive to the questions being asked in the survey. The paper concludes with an explanation of their outpatient experience index project (O-EXI) and the particular type of analysis (fuzzy set qualitative comparative analysis) that was used (and why).

7. Lorissa MacAllister

Dr. Lorissa MacAllister is a researcher, architect, founder, and President of Enviah a consulting company in healthcare that uses a pioneering approach to aligning operations and building environments around an organizational strategy for success. Her work applies research to practice and revolves around using the environment to improve occupants' experiences.

In Lorissa MacAllister's paper, she tackles the complex idea of brand value and how it can be defined and quantified. She discusses the ways in which qualitative research can be used to document the patient's experience as a series of encounters with healthcare professionals and the healthcare environment itself. Relying on a framework provided by Kahneman, the paper helps the reader understand that the patient's healthcare journey is comprised of a series of peak moments of encounter, such as when a nurse enters the patient's room. The compilation of these peak moments in turn creates a perspective about the quality of care the patient experienced. Research can help to document which encounters deserve more attention in building brand value.

8. Ellen Taylor

Ellen Taylor is the Vice President for Research at The Center for Health Design. She has a BArch from Cornell University, Global Executive MBAs from Columbia University and London Business School, and a PhD from Loughborough University in England. She was recognized as the HCD10 Researcher in 2017.

The final paper -- the values of secondary research -- essentially provides a bookend to the first paper on tightly controlled laboratory research. This paper by Taylor on the role of secondary research in helping the practitioner speaks to the realities of time pressures in the design field and what we can expect practitioners to digest, given the point Taylor makes that "rigorous research is being published at a rate that no one can realistically follow." The paper is laid out to present a challenge, such as "Time Searching" with a matched opportunity, in this case "Summarizing Evidence." There are five such challenge-opportunity sections in the paper, and in illustrating the solutions or opportunities, Taylor makes use of various tools developed by the Center for Health Design, such as the interactive design diagrams available on the Center website. The paper contextualizes secondary research in a way that challenges its connotation for academics as a shortcut of questionable value; readers will come away appreciating the ways in which secondary research provides opportunities for not only healthcare designers and practitioners, but for academics as well.

The Small Experiment: How Less is More in Healthcare Environment Research

Ann Sloan Devlin, *PhD (Connecticut College)*

ABSTRACT

The tension between tightly controlled experimental research and ecological validity is long-standing (e.g., Banaji & Crowder, 1989). Measured on any dimension, field research in healthcare environments is expensive. Before moving into the field, small, tightly controlled experimental research can provide evidence about which healthcare design interventions may be advantageous for healthcare outcomes, setting the stage to test the interventions in situ. Further, many researchers do not have either the connections to or support from healthcare administrators to conduct field research in healthcare settings. This presentation provides information about how to advance the knowledge base of healthcare environments in the absence of connections to a facility and on a small budget.

Drawing on a body of research dealing with inpatient and outpatient settings, the presenter will discuss approaches to conducting research in the lab with the potential for generalizability to healthcare environments. Examples will emphasize the benefits and drawbacks in three major areas: research design, visual stimuli, and participants. In research design, examples will include between subjects, within subjects, mixed design, and mixed methods. Regarding the creation of visual stimuli, approaches demonstrated in published work by the presenter will include taking photographs of real-world settings and using them directly, taking photographs of real world settings and manipulating them (e.g., in Photoshop), taking photographs of staged environments, and acquiring permission to use photographs and/or materials (e.g., posters) owned by others and related copyright issues. Participant issues will explore the use of participant pools, crowdsourcing platforms such as AmazonMTurk and Prolific, and intercept approaches.

In 1996, Rachel Kaplan, a methodological pioneer, published an article in the EDRA Proceedings focusing on what can be learned from small experiments. This session returns to focus on the benefits (and challenges) of doing laboratory work with implications for real word applications in healthcare.

INTRODUCTION

There is no perfect study that answers a research question completely, but a project, if well-designed, can answer some aspect of that question. Rachel Kaplan, a leader in the field of environmental psychology, stated that the project needs to be "about something—not everything" (1966, p. 172). Moreover, in the research that Stephen and Rachel Kaplan did as a team over five decades beginning in the 1960s, they created a model of how research advances through accretion—that is, through a gradual accumulation of layers. The small experiment can play a critical role in that process by identifying variables worth investigating further. And by examining variables in a tightly controlled manner, which is possible in an experiment, the researcher has some claim to causality for those variables.

Every study is a compromise between the tensions that exist in research—specifically the tight controls needed to contribute to claims of causality, and the extent to which the study approximates the real world— the question of the study's ecological validity. Researchers Mazarin Banaji and Robert Crowder described this tension in an article dealing with the phenomenon of everyday memory (Banaji

 https://doi.org/10.52202/067871-0001

& Crowder, 1989), which has been cited over 200 times (PsycINFO, Scopus), but the tension applies to all domains of research, including healthcare. Clearly, no one wants a study that has neither tight control nor ecological validity, and everyone desires a study that has both of those qualities. The focus here is research that combines tight control and "lower" (less than ideal) ecological validity. The goal of this paper is to demonstrate how to create greater ecological validity in the context of a small experiment.

Qualities of the "Small" Experiment

What is the "small" experiment? In this paper, the label is being used to describe a study, typically online, in which relatively few (i.e., 1-2) variables are manipulated using relatively few participants (< 300), often obtained through convenience, such as a university participant pool or crowdsourcing platform such as AmazonMTurk. Here, the examples used to illustrate the small experiment have in common being administered online with an emphasis on visual stimuli.

Advantages to the small experiment are numerous and include its 1) control (and the related concept of internal validity, which is the extent to which the research design allows you to adequately test the hypotheses, and specifically that the causal relationship you assess cannot be attributed to other factors); 2) time-savings—in contrast to field studies, the collection of data is generally rapid; 3) monetary savings (also in contrast to field studies)--fewer people are often needed to run the small experiment and collect the data, which is a financial savings if the research assistants are paid; 4) future-oriented contribution —because variables are tightly controlled, there are implications for causality, and the results of such studies can funnel into larger studies that involve the investment of greater capital and human resources. In other words, this kind of small project points to the variables that might be worthy of additional studies, moving from the laboratory into the field. For example, the Andrade and Devlin (2015) scenario study of responses to different combinations of the variables in Ulrich's Theory of Supportive Design (1991) led to a field study that tested those variables in real hospital settings with patients (Andrade et al., 2017).

Within the tight control of the experiment, there are ecological implications for both "the people" and "the stimuli." The participants can be selected to have qualities or experiences germane to the study (such as being actual patients or having visited a doctor's office) and the stimuli (creating the conditions to be manipulated) can come from a variety of domains that reflect the real (here healthcare) world, from those that document the real world (e.g., photographs of healthcare settings) to those that are generated artificially (e.g., waiting room seating arrangements created through software such as SketchUp).

Commonly Used Experimental Designs in Healthcare Research

The research design selected must be appropriate for the question to be answered. A few research designs commonly used in healthcare experiments will be briefly considered. In describing those designs, it is important to understand why the number of independent and dependent variables varies by design. Earlier I spoke about the tension or tradeoff between tight control and ecological validity. Tradeoffs also appear in terms of the number of IVs vs. DVs. Independent variables are typically manipulated (such as the number of credentials displayed on a therapist's wall; Devlin et al., 2009) or selected to be evaluated (e.g., Nasar & Devlin, 2011). Dependent variables are the outcomes measured, such as the quality of care expected, as a result of the exposure to the independent variable(s).

The tradeoff can be conceptualized as depth vs. breadth, which may be influenced by a very practical concern, the energy level of your participants. With only a limited amount of energy or willingness on their part to answer questions, you typically can ask more questions about fewer things (depth), or fewer questions about more things (breadth). As an example of depth, in a study involving the size of the artwork displayed in an exam room (Devlin et al., 2020) people saw one photograph—-but they answered quite a few questions about it in terms of the DVs: positive distraction (4 items), room spaciousness (4 items), anxiety (21 items), clinic evaluation (17 items), qualities of the physician (12 items), qualities of the room (4

https://doi.org/10.52202/067871-0001

items), as well as demographic questions. In contrast, Nasar and Devlin (2011; Study 1), examined more things with fewer questions (breadth). Every participant rated 30 pictures of different therapists' chairs but answered only two questions for each picture, the quality of care expected and how comfortable they would feel in it.

Between-Subjects Design (BSD)--the between-subjects design, in which different conditions of the experiment are randomly distributed across participants is frequently used. This approach has real advantages in experiments by isolating levels/versions of the variables that are manipulated to reveal their impact. BSDs have the advantage of not showing or exposing all versions of the stimuli to the same participants, which would certainly provide a "clue" about the purpose of the research and might lead participants to answer in a way they thought would support the hypothesis. The BSD can add complexity easily, by creating more "levels" or versions of the variable. For example, in a study examining the impact of the number of diplomas or credentials that were displayed in a psychotherapist's office on the judgments about the therapist, the research had four different levels of the variable, the number of credentials, in which 0, 2, 4, or 9 credentials were displayed in a photograph (with a given participant seeing only one of these levels; Devlin et al., 2009). Many researchers create complex (and informative) projects using only two variables, for example, 1) the size of the artwork displayed in an exam room (no image, small, medium, or large-size artwork) and 2) the amount of time that people are told they had to wait for the physician, either 10 or 45 minutes (Devlin et al., 2020). Using three or more variables would lead to complex interactions that are difficult to interpret. Complexity can be added to a study, with a concomitant increase in knowledge gained, by adding to the number of levels, without adding additional variables.

Within-Subjects Design (WSD)--in a within-subjects design, participants are exposed to all versions of the stimuli. Such exposure might lead people to make judgments different than if they were exposed to a single version. This kind of design has many uses in domains such as education where carryover (knowledge from the first exposure carries over to influence knowledge of the next exposure) is the focus of study. WSD is less frequently used in healthcare research but is applicable. Specifically, researchers often gather many examples of the environment (e.g., waiting rooms, therapy offices, hospital facades) and ask people to rate these examples on some outcome variable, like the quality of care expected. Participants necessarily are exposed to these stimuli in some order, so there is the potential for carryover effects. This is a case of within-subjects design because people are rating all the stimuli; often a technique like a factor analysis is used to reduce the number of stimuli (e.g., 30) to a small number of dimensions that share something in common. The potential impact of carryover is handled in various ways, such as creating different orders of the stimuli that people rate. Including different orders leads to a slightly different design, called the Mixed Design, which includes both a within and a between-subjects component.

Pre-Post Design, and Occupancy Design--the Pre-Post design reflects the idea of measuring something at Time 1 (e.g., before a renovation) and again at Time 2 (e.g., after a renovation) to look at the differences. However, without including a control group, you are not justified in labeling this an experiment. Measuring before and after an intervention is used in healthcare research; often the intervention is a renovation of a building or even a room. More commonly people just measure after the intervention; this approach has been called "Post-Occupancy" Evaluation or just "Occupancy Evaluation," to see how people react to the current situation. From the standpoint of claiming causality, there are numerous problems with a pre-post design without a control and without random assignment to the condition; you cannot attribute any of the differences to the intervention without comparing the outcome measures to a group that did not experience the intervention. Other problems involve the participants, who may not be the same at Time 1 and Time 2 and almost surely have not been randomly assigned to their condition (either experimental or control). Sometimes researchers approximate the pre-post design by showing the same people at the same time (WSD) or different people at the same time (BSD) pictures of the "before and after" condition. Figure 1 shows an example when a building corridor has been renovated with wayfinding aids.

 https://doi.org/10.52202/067871-0001

Figure 1

Pre- post Design

Unrenovated Renovated

Given these drawbacks, should researchers do non-experimental Occupancy Evaluations? My view is that doing brief interviews (e.g., about what works well, what does not) with a random selection of occupants of a renovated building or room or of new construction can lead to useful knowledge to be funneled into experimental research in the future.

Approaches to Visual Stimuli: Real, Real with Manipulations, Simulated with Real Elements, Graphics and Illustrations

The researcher can take steps to increase the ecological validity of an experiment through the stimuli and/or participants selected. Here, the stimuli discussed are visual elements, which are common in healthcare research, although other sensory modalities (e.g., auditory) might be used. I have categorized the visual stimuli on a dimension that varies from real to depictions using illustrations or 3-D modeling, in this case meaning stimuli that are photographs of real environments (real) to stimuli that are generated by the researcher (illustrations; graphics from tools like SketchUp).

Photographs from the Real World

Exposing people to one (BSD) or more (WSD) photographs of actual environments advances ecological validity because these images exist in the real world. The Kaplans often used this approach in their work on natural environments (see Kaplan & Kaplan, 1989), and colleagues and I have used this approach in several studies (see, for example, Arneill & Devlin, 2002; Noble & Devlin, 2021). In a study of medical office exteriors (Devlin, 2008), photographs of the facades of 34 medical office facilities in Michigan, Rhode Island, and Connecticut were taken from public access standpoints to capture the exteriors of the structures (see Figure 2). Where the space resides on a public-private continuum will determine the nature of the permission(s) involved in acquiring these pictures. To use photographs of interiors of privately owned or rented spaces, written permission is necessary. Photographs of real environments also exist in the form of published works, such as books, and permission to reuse such copyrighted material must be obtained.

https://doi.org/10.52202/067871-0001

Figure 2
Medical Office Exteriors

Photographs from the Real World with Manipulations

In some cases, to test the variable of interest, a photograph of a real environment may need to be changed in a systematic way, for example though use of a software graphics editor. An example is the study by Devlin et al. (2020), in which the original photograph (marked by an "x" in Figure 3 and considered the small image) was edited to produce three additional versions: no artwork on the wall, a medium-sized version, and a large version. This editing allowed the researchers to systematically test the impact of different versions of having artwork on a wall, on participants' reactions to such judgments as the quality of care they expected to receive. In a more elaborate change involving a "real environment," the number of credentials displayed on a therapist's wall (Devlin et al., 2009) was varied in the therapist's office (affixed with removable mounting tape so as not to damage the wall). The researchers began with a therapy office (offered for use by a colleague of the lead researcher) and modified the environment to create four levels of the credentials variable: 0, 2, 4, or 9 credentials. Photographs of each version were used in the research.

Figure 3
Manipulation of Artwork Size

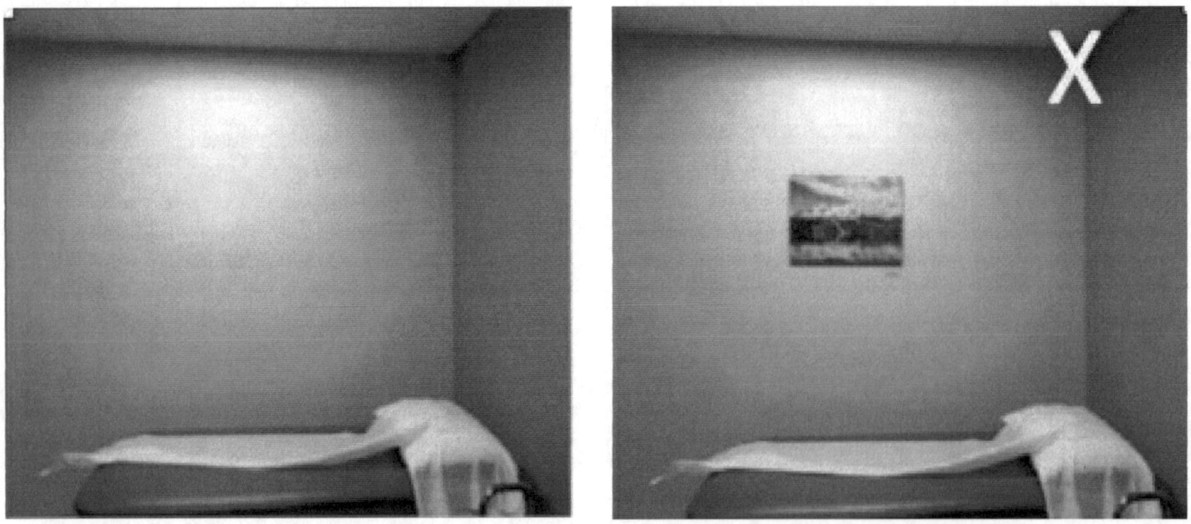

Blank Wall Small Image

 https://doi.org/10.52202/067871-0001

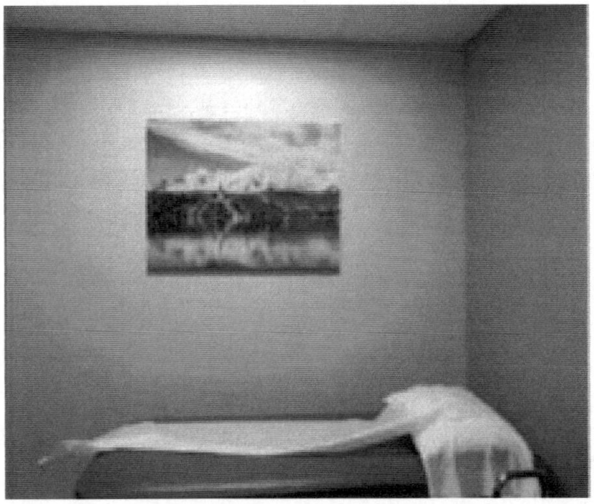

Medium Image

Large Image

(From Devlin et al., 2020)

Simulation of the Environment with Real Elements

Where no environment exists to test the variables of interest, you may need to create it yourself. In one such study (Devlin et al., 2014), on consultation spaces, we examined the mode of information delivery (paper vs. computer), table shape (curved vs. rectilinear) and seating arrangement (side-by-side vs. opposite). To create this consultation space, we re-purposed a lab room and used some available furniture. We then took photographs of the eight conditions (examples shown in Figure 4).

Figure 4

Simulation of a Consultation Space

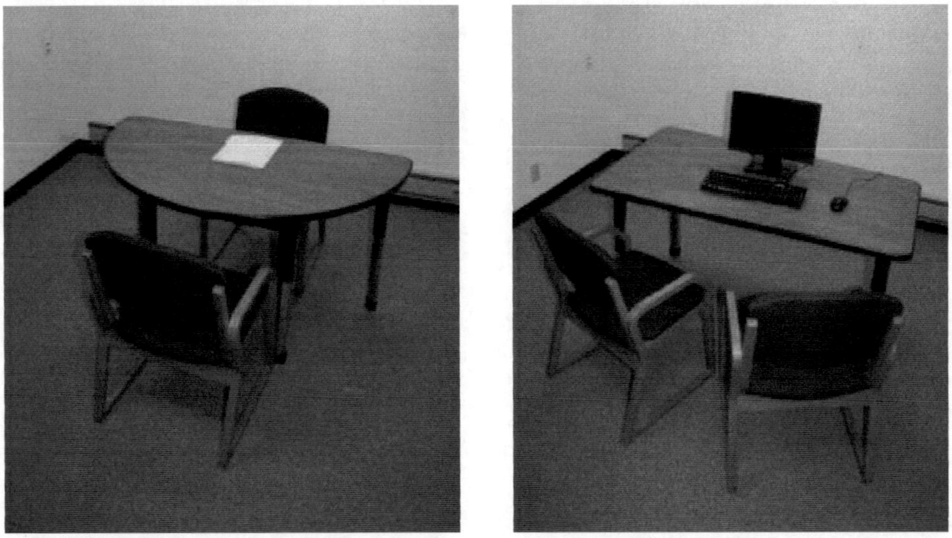

Experimenter-created Stimuli

A final category is stimuli created by the researcher such as an illustration or using a 3D modeling application. In a recent study (Devlin, 2022), five different seating arrangements were created through SketchUp and used in a BSD to gather participants' preferences for those seating options. In another example (see Figure 5), an artist and friend of the researcher created an illustration of an exam room for a study about preferred physician (P) and medical scribe (S) location in that room.

https://doi.org/10.52202/067871-0001

Figure 5

Illustration of an Exam Room

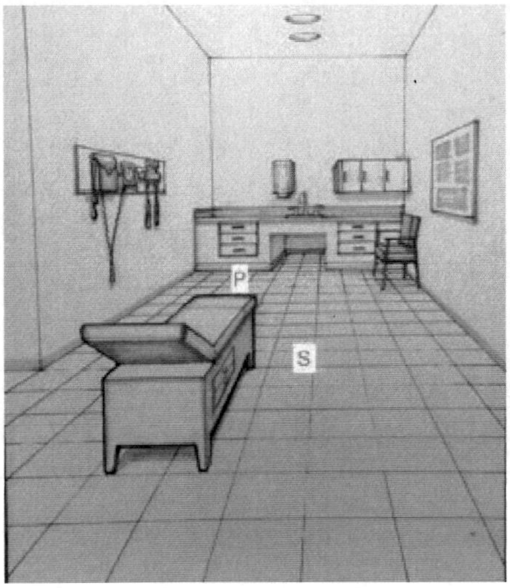

A Word About Using Photographs

One criticism raised about using photographs is that on-site responses may differ from those given to photographs. Some literature, e.g., a meta-analysis from Stamps (1993), suggests a high level of correlation between responses gathered on site and those captured in color photographs of the same environment. Use of iconic simulation such as photographs has been argued to be a valid research strategy (Ulrich et al., 1991).

Participants in Online Studies: University Pools and Crowdsourcing Platforms Who participates in your study can contribute to its ecological validity. Two categories of participants available for online studies (the focus here) are university participant pools and crowdsourcing platforms.

University Participant Pools

University participant pools typically consist of individuals (typically undergraduates) at a college or university. Such pools are accessible and free, although regulations often exist about how many participants one may use from the pool each semester. Despite the numerous advantages of such pools, a disadvantage is their limited size. Gaining the requisite number of participants needed for such approaches as factor analysis, especially in a single semester, might be a challenge.

Additional drawbacks may create threats to your ecological validity, including homogeneous participant characteristics and lack representativeness of the broader population. My opinion is that using students as participants is appropriate for studies in which students could be expected to have relevant experience or familiarity with the situation being examined. In my studies on psychotherapy offices, over 50% of the participants reported at least one visit to a therapist. The percentage of those with experience is even higher when the question is about simply visiting a doctor's office. Thus, in the context of healthcare, many if not most students have relevant experience with the outpatient setting.

Crowdsourcing Platforms

Crowdsourcing platforms, such as Amazon Mechanical Turk (MTurk) are sources for participants (workers) who are paid small amounts of money (often under $1) to respond to surveys or other human intelligence tasks (HITs) as they are known. There has been a large increase in the popularity of such sites, for example an over 2000% increase in use from 2012-2019 in management studies (Aguinis et al., 2021). Such platforms have numerous advantages, including the size of the participant pool and the speed of

 https://doi.org/10.52202/067871-0001

collection data (~300 participants in less than 3 hours, in my experience). Some researchers have claimed that the quality of the data are as good, if not better, than those of participant pools in many domains — social, clinical, cognitive, and personality (Buhrmester et al., 2011; Chandler et al., 2014). Using MTurk is also relatively inexpensive, but there is still payment.

What are the disadvantages to using MTurk? Primary among them in my view is the issue of data quality. The article by Aguinis et al. (2021; see Table 2 in that article) lists a number of problems, including inattention by workers, self-misrepresentation (such as saying they represent a particular demographic), self-selection bias, high attrition rates, inconsistent English language fluency, non-naïveté, growth of MTurker communities (so that they interact with one another and may communicate the purpose of a given study), vulnerability to Web bots, social desirability bias, perceived researcher unfairness (e.g., to make compensation decisions or lack of ability to communicate with the researcher). Of these disadvantages, a number would be true of participant pools as well, such as inattention, social desirability bias, and interaction (students talk to each other about the projects).

In my experience, the largest disadvantage is inattention. As an example, in 2013, for a study I did, 62.7% of the MTurk participants passed both manipulation checks. When the study was repeated in 2021 to look at seating arrangements after the onset of the pandemic, 45.7% passed the manipulation checks. Agley et al. (2021, August 6) suggest control questions you can incorporate to help manage data quality (and remove participants who do not pass the questions).

Other crowdsourcing sites such as Prolific (www.prolific.com) have emerged as an alternative to MTurk; Prolific was designed for researchers. The advantages to that platform, according to their materials, include "highly reliable data," and ethical rewards (a minimum hourly rate), among others. A recent study (Peer et al., 2017) found MTurk and Prolific comparable on data quality. Literature in health and medical research concludes that MTurk is an "efficient, reliable, cost-effective tool" for generating responses comparable to those collected through more conventional means (Morentsen & Hughes, 2018, p. 533).

Ways to Improve Data Quality

There are numerous ways to improve the quality of data collected through platforms like Amazon MTurk. A number of these suggestions are part of what makes something a good experiment, in particular the idea of standardization and a systematic approach to data collection. One suggestion, especially for use with photographs, is to use a timing mechanism, which survey softwares such as Qualtrics support. A timing mechanism enables you to keep participants on a page for a given length of time, say 15 sec. No, you cannot guarantee that they will look at the photograph for the full 15 sec, but at least they will not be able to advance the survey until 15 sec have elapsed. You can measure the length of time people take for the entire survey, and eliminate people who "rush through," whatever your operational definition of rushing is for your study. You can set parameters within the AmazonMTurk platform, such as country in which the participant resides and the approval rating the participant has received from work on other tasks (e.g., 97%).

Quality assurance mechanisms like manipulation checks and attention checks should be a standard part of the survey. A manipulation check assesses whether participants can recall or identify some aspect of the study that was manipulated, such as which one of the four photographs of art in an exam room (no art, small image, medium image, large image) they saw. An attention check tests whether people are reading carefully. An example would be to include a question such as "Check response option #3 in this question." If people fail attention and manipulation checks, researchers often omit them from formal analyses, although practices for this vary.

Conclusion

Small experiments provide a wide variety of opportunities in research design, stimuli, and participants to improve healthcare environments. Such research can help researchers decide what variables are worth studying on a larger scale and in the field. The small experiment and the approaches that support it are particularly useful for researchers at small schools with limited resources.

https://doi.org/10.52202/067871-0001

References

Agley, J., Xiao, Y., Noland, R., & Golzarri-Arroyo, L. (2021, August 6). Quality control questions on Amazon's Mechanical Turk (MTurk): A randomized trial of impact on the USAUDIT, PHQ-9, and GAD-7. *Behavior Research Methods, 54*, 885-897 (2022). https://doi.org/10.3758/s13428-021-01665-8

Aguinis, H., Villamor, I., & Romani, R. S. (2021). MTurk research: Review and recommendations. *Journal of Management, 47* (4), 823-837. https://doi.org/10.1177/0149206320969787

Andrade, C., & Devlin, A. S. (2015). Stress reduction in the hospital room: Applying Ulrich's theory of supportive design. *Journal of Environmental Psychology, 41*, 125-134. https://doi.org/10.1016/j.jenvp.2014.12.001

Andrade, C. C., Devlin, A. S., Pereira, C. P. & Lima, L. (2017). Do the hospital rooms make a difference for patients' stress? A multilevel analysis of the role of perceived control, positive distraction, and social support. *Journal of Environmental Psychology, 53*, 63-72. https://doi.org/1016/j.jenvp.2017.06.008

Arneill, A., & Devlin, A. S. (2002). Perceived quality of care: The influence of the waiting room environment. *Journal of Environmental Psychology, 22*, 345-360. https://doi.org/10.1006/jevp.2002.0274

Banaji, M. R., & Crowder, R. (1989). The bankruptcy of everyday memory. *America Psychologist, 44* (9)1185-1103 https://doi.org/10.1037/0003-0066X.44.9.1185

Buhrmester, M., Kwang, T., & Gosling, S. D. (2011). Amazon's Mechanical Turk: A new source of inexpensive, yet high-quality data? *Perspectives on Psychological Science, 6* (1), 3-5. https://doi.org/10.1177/1745691610393980

Chandler, J., Mueller, & P., Paolacci, G. (2014). Non-naivete among Amazon Mechanical Turk workers: Consequences and solutions for behavioral researchers. *Behavioral Research, 46* (1), 112-130. https://doi.org/10.3758/s13428-013-0365-7

Devlin, A. S. (2008). Judging a book by its cover: Medical building facades and judgments of care. *Environment and Behavior, 40*, 307-329. https://doi.org/10.1177/0013916507302242

Devlin, A. S (2022). Seating in doctors' waiting rooms: Has COVID-19 changed our choices? *Health Environments Research and Design Journal.* Advance online publication. https://doi.org/10.1177/19375867221104248

Devlin, A. S., Adler, J., Andrade, C., Browarnik, B., Evans, K., Finch, C., Plummer, G., & Watkins, K. (2014, May 29). *Consultation Spaces: Furnishings and Mode of Medical Information.* Poster presentation, 2014 Environmental Design Research Association Conference, New Orleans, Louisiana.

Devlin, A. S., Anderson, A., Hession-Kunz, S., Kelly, M., Noble, L., & Zou, A. (2020). Magnitude matters: Art image size and waiting time impact perceived quality of care. *Health Environments Research and Design Journal, 13*(3), 140-153. https://doi.org/10.1177/1937586719892602

Devlin, A. S., Donovan, S., Nicolov, A., Nold, O., Packard, A., & Zandan, G. (2009). "Impressive?" Credentials, family photographs, and the perception of therapist qualities. *Journal of Environmental Psychology, 29*, 503-512. https://doi.org/ 10.1016/j.jenvp.2009.08.008

Kaplan, R. (1996). The small experiment: Achieving more with less. In J. L. Nasar & B. B. Brown (Eds.), *Public and private places: Proceedings of the 27th annual Environmental Design Research Association* (pp. 170-174). EDRA.

Kaplan, R., & Kaplan, S. (1989). *The experience of nature: A psychological perspective*. Cambridge University Press.

Mortensen, K., & Hughes, T. L. (2018). Comparing Amazon's Mechanical Turk platform to conventional data collection methods in health and medical research literature. *Journal of General Internal Medicine, 33*(4), 533-538. https://doi.org/10.1007/s11606-017-4246-0

Nasar, J., & Devlin, A. S. (2011). Impressions of psychotherapists' offices. *Journal of Counseling Psychology, 58*, 310-320. https://doi.org/10.1037/a0023887

Noble, L., & Devlin, A. S. (2021). Perceptions of psychotherapy waiting rooms: Design recommendations. *Health Environments Research & Design Journal, 14*(3), 140-154. https://doi.

 https://doi.org/10.52202/067871-0001

org/10.1177/19375867211001885

Peer, E., Brandimarte, L., Samat, S., & Acquisti, A. (2017). Beyond the Turk: Alternative platforms for crowdsourcing behavioral research. *Journal of Experimental Social Psychology, 70*, 153-163. https://doi.org/10.1016/j.jesp.2017.01.006

Stamps, A. E. (1993). Simulation effects on environmental preference. *Journal of Environmental Management, 38*, 115–132. https://doi.org/10.1006/jema.1993. 1033

Ulrich, R. S. (1991). Effects of interior design on wellness. Theory and recent scientific research. *Journal of Health Care Interior Design, 3*(1), 97e109.

Ulrich, R. S., Simons, R. F., Losito, B. D., Fiorito, E., Miles, M. A., & Zelson, M. (1991). Stress recovery during exposure to natural and urban environments. *Journal of Environmental Psychology, 11*, 201–230. https://doi.org/10.1016/S0272-4944(05)880184-7

Finding Our Way in the Wayfinding Study: How Laboratory Work Influences a Randomized Controlled Trial

Rebecca Davis, *PhD, RN, FGSA (Grand Valley State University) and* **Margaret Calkins**, *PhD, EDAC, FGSA (IDEAS Institute)*

ABSTRACT

Methods to measure the impact of environmental features on wayfinding are challenging due to many factors, such as changing environmental conditions and lack of reproducibility. The Wayfinding Study team has worked for a number of years to develop sophisticated methods of measuring both the features of the environment and wayfinding ability. Initial studies conducted were aimed at measuring the types of environmental features, such as color, form, size, and placement, that impacted wayfinding ability. These studies were conducted in virtual reality environments in a laboratory. In addition, eye tracking technology was employed to measure visual fixations onto the environmental features, giving evidence for the types of features that are most engaging. Using results from these laboratory data, the Wayfinding Study Team developed an environmental intervention (special signage and art cues) aimed at improving wayfinding ability in persons with cognitive impairment who live in long term care settings. The NIA funded randomized controlled trial is being conducted in 12 real world settings. The team identifies three routes in each community (simple, moderate, and complex) and uses space syntax to measure the objective complexity of each route. A limitation of space syntax is that it does not account for wayfinding signage, lighting, or other environmental features other than connectedness within the building plan. The team has developed a new tool to account for the impact of these features (the TAWC; the Tool for Assessing Wayfinding Complexity). The team uses an innovative technology, DOTT, for measuring wayfinding performance. They also are using real time location tracking to determine if participants in sites receiving the intervention interact more within the community. The evolution of this research is an example of moving from laboratory to real world in the conduct of a RCT to measure the effectiveness of an environmental intervention for wayfinding design.

INTRODUCTION

Wayfinding, the ability to successfully navigate to a destination, is a known problem in persons with Alzheimer's disease and related disorders (ADRD), the most common cause of dementia (Alzheimer's Association, 2022). Getting lost in familiar and unfamiliar environments is among the earliest signs of the disease; and the problem progresses to affect almost all persons with ADRD (Allison et al., 2016). Long-term care (LTC) communities often have many long, virtually identical, and confusing hallways; however, the population housed in these communities is vulnerable to getting lost. Most older adults in LTC settings have some form of disability; the most prevalent is cognitive disability along with physical mobility problems. These LTC settings often have poor signage, a lack of visual cues, confusing floorplans, complex decision points, and other features that make them difficult for wayfinding (Marquardt, 2011).

Our research over the years has focused on methods to develop a wayfinding intervention for older adults in LTC settings. Based on the theories of cognitive maps and other cognitive science, we determined that an important method of environmental support for wayfinding is the presence of environmental cues. Environmental cues include both signage and landmarks, purposefully placed in key areas to provide information for wayfinding. Both signage and landmarks are necessary for wayfinding

 https://doi.org/10.52202/067871-0002

yet have different purposes (see Table 1). However, it has been unclear, based on current literature, what features of landmarks and signage best suited older adults; especially those with ADRD. The initial barriers to our research to develop wayfinding interventions included how to measure wayfinding ability in a frail population that often cannot walk very far (or at all). Also, we were unsure how to create equivalent conditions across participants to make valid inferences about our interventions, given that environments, by their very nature, change constantly in terms of light, sound, and activity. We needed to determine how to measure environmental features, so that they could be reproduced as interventions. We also needed to define the characteristics of landmarks such as size, color, shape, and meaningfulness; and to determine a method to measure these characteristics. The rest of the paper will describe the journey of this research team from developing interventions and testing methods from the laboratory to the real world and discuss the challenges and successes of these methods.

Table 1

Different Contributions of Landmarks and Signage for Wayfinding

Signage	Landmarks
Necessary for wayfinding in new environments	Necessary for environmental knowledge in new and familiar environments
Key for complex environments with many decision points	In complex environments, they support knowledge of location
For those with memory impairment, signage provides directions for concrete routes	Necessary for long term "knowing" of an environment for maximal flexibility in wayfinding
Can be seen close-up	Can be seen from a distance (if large)

Laboratory Testing

The first few studies the team conducted were based in a virtual reality (VR) platform so that the environment could be tightly controlled. The purpose of the first studies was to determine the properties of environmental landmarks that made them easier to use for wayfinding. The studies used a laptop computer program called The Computer Generated Arena (Nadel, et al., 1998). Older adults had to find a "hidden platform" in a round arena, which was only memorable based on visual landmarks placed on the wall. The landmark cues varied with respect to their features; color (versus black and white), size, number, and abstract versus nameable. Wayfinding was measured by recording the speed in which participants found the platform over repeated trials as well as the number of errors they made (Davis & Therrien, 2012; Davis et al., 2008, 2009). From these studies, results showed that older participants found their way to the platform, and remembered its location best, when cues were colorful and when they were simple, nameable objects.

In a subsequent VR study, a "Virtual Senior Living" (VSL) was developed to look like a real long term care community (Davis & Weisbeck, 2016). The lifelike rendition was shown on a 12-foot screen, and participants with and without Alzheimer's disease had to find their way in two similar environments that differed only with respect to the cues (see Figure 1). Participants wore eye tracking glasses that tracked what they visually attended to during active wayfinding. Cues that were nameable and colorful (like a rainbow, flag, balloons) were implemented in one environment at key decision points; the other environment was devoid of cues. Both persons with Alzheimer's disease and those without found their way better in the cued environment (Davis et al., 2017). Persons with Alzheimer's disease fixated less on meaningful environmental information than did those without the disease, which may account for some of their difficulty in wayfinding (Davis & Sikorskii, 2020).

https://doi.org/10.52202/067871-0002

Figure 1

Laboratory Testing Using the Virtual Senior Living Virtual Reality Program

Note: The VSL was projected on a 12-foot screen; participants found their way to a specific location using a joystick. Participants wore eye tracking glasses to track visual fixations while wayfinding (Davis et al., 2017). Photo credit: Michele Coffill, Grand Valley State University, University Communications.

Figure 2

The Virtual Senior Living Conditions

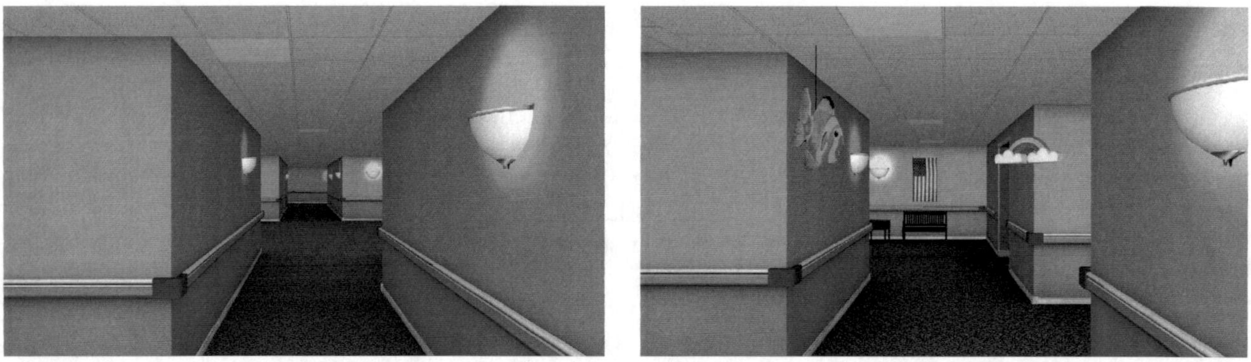

These are two virtual environments in which persons with and without Alzheimer's disease found their way; on the left shows a traditional senior environment with no signs or landmarks. Specially designed cues were placed at key decision points in the alternative environment on the right. Figure reproduced with permission (Davis et al., 2017).

From these studies, we learned that older adults with and without Alzheimer's disease found their way more often with colorful, real-life, nameable, landmarks than with black and white landmarks; lack of landmarks, or abstract landmarks. Big, colorful objects that stand out from the surroundings were most effective at assisting with wayfinding. Cues were especially important at decision points. People also use environmental features such as handrails and doors for wayfinding. Persons with Alzheimer's disease used fewer strategies and fixated less overall on cues while wayfinding compared to those without the diagnosis.

 https://doi.org/10.52202/067871-0002

We also found many pros and cons to using VR for testing wayfinding behavior. These pros and cons are listed in Table 2.

Table 2

Pros and Cons of Virtual Reality Testing

Criteria	Pros	Cons
Environmental Conditions	Can be the same for all participants	Is not "real"
Measurement	Can measure exactly using computerized techniques	May not capture the opinions and qualitative characteristics of experiencing the real environment
Comparison environments	Can have multiple environments the same length and same complexity	Definition of complexity is somewhat subjective
Mobility	Does not require the ability to walk, inclusive of many	Does not include any wayfinding challenges that occur with true physical mobility
Ease of use	Can be brought to individuals, easy to set up and use.	Computers may be unfamiliar to older adults and cause anxiety
Validity	Shown to be valid in many studies	Limited translation to real world
Use	First step in intervention development	Need to move to real world at some point
Side Effects	Newer methods of VR have less simulation sickness	A certain percent of participants experience simulation sickness

Real World Testing

Currently, the research team is conducting a real-world wayfinding study (The Wayfinding Study) using the findings from the laboratory studies. The study is an NIA funded cluster, randomized controlled trial (R01AG056384), conducted in 15 long term care (LTC) settings (independent and assisted living residences). The aim of the study is to assess the impact of age-friendly signage, art cues, and spaced-retrieval education on wayfinding in persons with cognitive impairment who live in long-term care communities. Spaced-retrieval is a memory-training technique where retrieval of information is practiced at increasingly longer intervals of time (Brush & Camp, 1998).

The study has three arms, including 1) control (no intervention); 2) art cues and signage added; and 3) art cues, signage, and spaced retrieval education added. Figure 3 shows the signage and art cue pre and post-intervention and Figure 4 shows an example of the signage. The art cues were chosen based on prior studies and are large, easily identifiable, objects that are distinctive and if possible, represent the area in which they are placed (salient).

Wayfinding was assessed by asking participants to find their way to designated locations on three routes – simple, moderate, and complex. The routes were chosen by the researchers based on predetermined criteria. For example, the complex route most often involved a change in floors (elevator) if possible. Space syntax (discussed elsewhere by Saif Haq in these proceedings) was used to assess the complexity of each route by looking at the R3 (spatial integration) value for each route. Complex routes had less integration than did the moderate routes, which in turn had less integration than the simple routes. Participants were timed as they traversed the route, and the errors made were documented at the location using a computer program on an iPad.

Figure 3

The Wayfinding Study Intervention

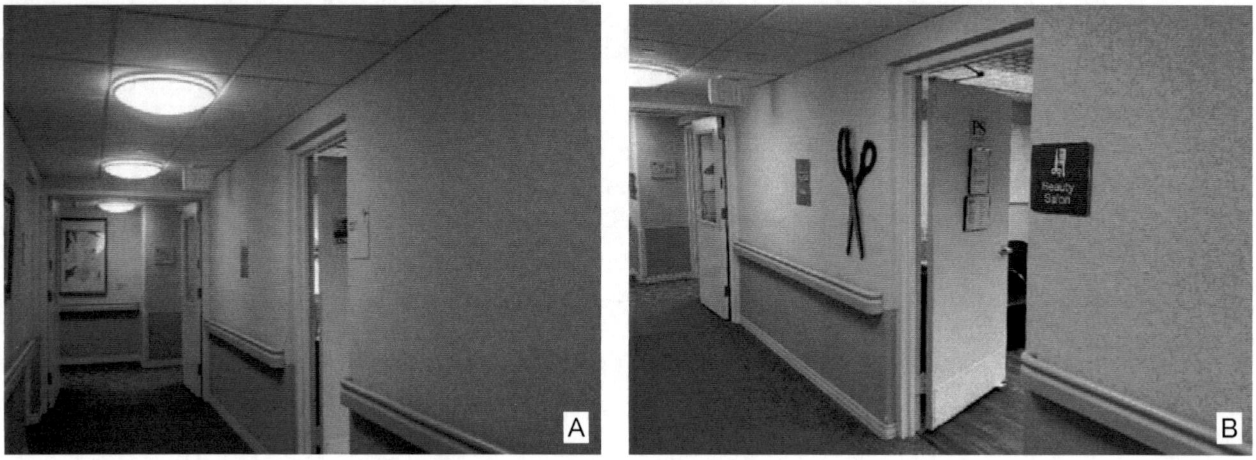

Note: Figure A is a picture of the beauty salon prior to intervention. Figure B is the same location after the cue (scissors) and new dementia-friendly signage has been added.

Figure 4

Age and Dementia Friendly Signage

Note: Age and dementia friendly signage included high contrast letters and background; large upper and lower- case letters; and icons.

Measurement in Real World Environments

Real world testing has many challenges. It is necessary to have a measure of environmental complexity prior to the intervention. Factors such as the complexity of the decision points, the existing environmental design, and the length of routes, among others, are factors that must be systematically measured to compare environments and generalize the results.

We utilized three methods to assess environmental complexity. These include space syntax (R3

 https://doi.org/10.52202/067871-0002

Axial Integration) (Zook, 2017); a modified version of The Wayfinding Checklist (Benbow, 2013); and a newly devised tool (developed by this research team) called the Tool for Assessing Wayfinding Complexity (TAWC).

As mentioned earlier, the R3 values are provided for each individual route, as well as the LTC community as a whole. These values are calculated using space syntax methods (Zook, 2017). The Wayfinding Checklist is a simple-to-use checklist, giving values for common wayfinding design features such as the floorplan (hallway length, etc.) and design elements such as lighting and signage (Benbow, 2013). The checklist is applicable to a whole building, not a specific route. The TAWC was developed due to a lack of existing methodology to assess specific routes for complexity other than space syntax – especially the visual field elements such as the presence of signage and landmarks. The tool has two components; a floor plan analysis, which gives values for the types of decision points in a route; and a visual field analysis, which examines specific design features for each decision point along a route. The TAWC gives researchers the ability to assess design features along routes before and after an intervention. The tools each have pros and cons, which are listed in Table 3.

There are many benefits to real world testing. It is necessary to conduct real world tests to ensure the effectiveness of interventions and demonstrate their ecological validity. However, many factors can affect the generalizability of the study results. Good measures of environmental complexity can improve the ability to generalize results. Taking into consideration threats to external validity, such as historical events (e.g., COVID restrictions that impaired resident mobility for over a year) that might impact wayfinding ability, are also important.

In summary, we have discussed the evolution of developing and testing an intervention to assist persons with cognitive impairment find their way in complex environments. Both laboratory methods, such as virtual reality, and real-life testing have challenges and benefits; and both are useful in the development of an intervention.

Table 3

Pros and Cons of Three Types of Environmental Measures for Wayfinding Design

Tools	**Pros**	**Cons**
Space syntax	Scientifically based; well-known and systematic; valid measure of floorplan. Can be used to measure routes and buildings. Provides interval level data.	Does not account for visual field. Values do not always match with perceived complexity of a space. Cannot be used to measure pre/post signage and cues intervention.
Wayfinding Checklist	Overall measure of floorplan and visual field design features. Can be used to compare overall complexity between buildings. Includes common features related to wayfinding ability.	No validity testing. Does not measure differences between routes. Limited ability to assess before/after design intervention.
TAWC	Includes measures of visual field and floorplan complexity. Measures complexity of routes. Can be used to measure before/after intervention. Provides interval level data.	Takes time to evaluate all aspects of a route. Takes some knowledge of floorplans. Currently needs validity and reliability testing.

https://doi.org/10.52202/067871-0002

References

Allison, S. L., Fagan, A. M., Morris, J. C., & Head, D. (2016). Spatial navigation in preclinical Alzheimer's Disease. *Journal of Alzheimer's Disease, 52*(1), 77-90. https://doi.org/10.3233/JAD-150855

Alzheimer's Association. (2022). 2022 Alzheimer's disease facts and figures. *Alzheimer's & Dementia, 18*(4), 700-789. https://doi.org/https://doi.org/10.1002/alz.12638

Benbow, W. (2013). Evidence-based checklist for wayfinding design in dementia care facilities. *Canadian Nursing Home, 24*(1), 4-10.

Brush, J. A., & Camp, C. J. (1998). Using spaced retrieval as an intervention during speech- language therapy. *Clinical Gerontologist, 19*(1), 51-64.

Davis, R., Ohman, J. M., & Weisbeck, C. (2017). Salient cues and wayfinding in Alzheimer's disease within a virtual senior residence. *Environment and Behavior, 49*(9), 1038-1065. https://doi.org/10.1177/0013916516677341

Davis, R., & Sikorskii, A. (2020). Eye tracking analsyis of visual cues during wayfinding in early stage Alzheimer's disease. *Dementia and Geriatric Cognitive Disorders, 49*, 91-97. https://doi.org/https://doi.org/10.1159/000506859

Davis, R., & Weisbeck, C. (2016). Creating a supportive environment using cues for wayfinding in dementia. *Journal of Gerontological Nursing, 42*(3), 36-44. https://doi.org/10.3928/00989134-20160212-07

Davis, R. L., & Therrien, B. A. (2012). Cue color and familiarity in place learning for older adults. *Research in Gerontological Nursing, 5*(2), 138-148. https://doi.org/10.3928/19404921-20111004-01

Davis, R. L., Therrien, B. A., & West, B. T. (2008). Cue conditions and wayfinding in older and younger women. *Research in Gerontological Nursing, 1*(4), 252-263. https://doi.org/10.3928/19404921-20081001-06

Davis, R. L., Therrien, B. A., & West, B. T. (2009). Working memory, cues, and wayfinding in older women. *Journal of Applied Gerontology, 28*(6), 743-767. https://doi.org/10.1177/0733464809332785

Marquardt, G. (2011). Wayfinding for people with dementia: A review of the role of architectural design. *Health Environment Research & Design, 4*(2), 75-90. https://doi.org/10.1177/193758671100400207

Nadel, L., Thomas, K. G., Laurance, H. E., Skelton, R., Tal, T., & Jacobs, W. J. (1998). Human place learning in a computer generated arena. *In Spatial cognition (pp. 399-427).* Springer, Berlin, Heidelberg.

Zook, J. B. (2017). Spatial schemata in museum floorplans. *Proceedings of the 11th International Space Syntax Symposium. 11th International Space Syntax Symposium,* Lisbon, Spain.

 https://doi.org/10.52202/067871-0002

Strategic Prioritization of an Occupancy Evaluation Program for Healthcare Design

***Kara Freihoefer**, PhD, NCIDQ, EDAC, LEED ID+C and **Renae Rich**, MS, EDAC (HGA Architects and Engineers)*

ABSTRACT

Traditionally known as post-occupancy evaluation, a holistic occupancy evaluation serves the dual intent of collecting data both pre- and post-occupancy of a particular healthcare design project. The benefit of gathering pre-occupancy data is two-fold; it provides insights on current-state, baseline measures and if gathered during the pre-design phases of a project process, insights can be applied to inform the future-state environment. Then, once constructed and operated in for at least six-months, the same protocol, methods, and tools can be used to gather post-occupancy data. The benefit of having both datasets allows for analyses around the magnitude of change because of the newly designed environment. However, because all design variables changed in construction, attributing findings to one design solution is impossible.

Regardless, there is still immense value in gathering insights and data, pre- and post-occupancy on a project-by-project basis, but with standardization of protocols, methods, and tools a robust and sound comprehensive occupancy evaluation program could provide opportunities for meta-analyses. A comprehensive occupancy evaluation also could encourage translational application of findings and discovery of relationships among key design interventions, all while further advancing the role of design research in practice.

This presentation will build the case for the development of a holistic occupancy evaluation program for healthcare settings, discuss a framework for protocol and tool development that engages design practitioners, review opportunities for standardization among various health facility types and across architecture industries, demonstrate data analyses on a meta-analyses scale, and highlight constructs of a data management system. The presentation will feature strategies, obstacles, frameworks, and limitations in the development of an in-house comprehensive occupancy evaluation program and include discussion on extending a collaborative effort among institutions and firms.

INTRODUCTION

Precedent-setting literature and research on the built environment, published over the last 60+ years, has compelled architecture and engineering firms to prioritize occupancy evaluations as a strategic initiative. The initiative has led some organizations to develop standardized assessment tools and procedures to evaluate similar built environments.

HGA Architects and Engineers is among the firms embracing such an approach. Recently HGA's executive leadership established a long-term strategic plan consisting of three foundational pillars (see Figure 1).

https://doi.org/10.52202/067871-0003

Figure 1
HGA Three Strategic Pillars

The first pillar in the strategic plan— "to build the value of the firm as a curious, research-driven organization"—demonstrates HGA's commitment to using data, evidence, and original insights to inform design. Integral to this pillar is a stated goal to garner a reputation for research and measured results. To help achieve this goal, HGA launched an initiative to "implement a scalable, pre- and post-occupancy evaluation program across all market sectors of the firm."

The intent of this paper is to provide historical precedent as it relates to occupancy evaluations (possible at any time, not only post-occupancy), outline the evolution, examine barriers to implementation in practice, and detail how, as a research-driven organization, HGA is developing a robust occupancy evaluation program to prioritize and strategically align recurrent and consistent evaluation of the built environment.

Occupancy Evaluation History and Evolution

Some of the first instances of feedback about the built environment were precipitated by major building failures and resulted in construction regulations and building codes. However, through connections with psychology and sociology, the practice of evaluating completed buildings from the occupant perspective arose from isolated case studies in the late 1960s and progressed to include system-wide and cross-sectional evaluation efforts by the 1970s and 1980s (Preiser et al., 1988). Since then, many have cited the following Preiser et al. (1988, p. 3) definition: "Post-occupancy evaluation (POE) is the process of evaluating buildings in a systematic and rigorous manner after they have been built and occupied for some time."

This brief description has since been widely adopted and implemented to assess completed buildings, but it falls short in addressing the breadth of information and insights occupancy evaluations have to offer. In fact, this is not where the Preiser definition ended, and the subsequent text is arguably more vital to a successful evaluation: "...POEs focus on building occupants and their needs, and thus they provide insights into the consequences of past design decisions and the resulting building performance. This knowledge forms a sound basis for creating better buildings in the future" (Preiser et al., 1988, p. 3).

Unfortunately, the declaration that evaluation should happen after facilities were built and occupied, coupled with the term POE itself, made certain that occupancy evaluations were destined to be an afterthought in the design process, a separate endeavor often overlooked and undervalued as a key player throughout and between design projects.

The Value of Occupancy Evaluation

A well-planned and appropriate occupancy evaluation, performed at any phase in the design process (not just post-occupancy), offers much value to a variety of stakeholders: design teams, facilities

 https://doi.org/10.52202/067871-0003

managers, organization leaders, occupants themselves, and the industry at large, to name a few. An evaluation may support an organization and its members in any of the following ways, adapted from Parshall (1989):

- Justify actions or expenditures, particularly as accountability for large or public investments
- Assess finished quality and whether the design meets expectations
- Inform iterative designs in future building phases or similar facility types at the same or other organizations
- Test application of new or innovative design solutions
- Understand current functionality and make adjustments for optimal effectiveness and use
- Provide opportunities for feedback from and education of occupants

Similarly, Van der Ryn and Silverstein suggested that evaluations of built environments serve four functions: compare actual to intended use (evaluative), provide organized information to designers to reduce uncertainty (informative), create opportunities for innovation through connections between form and function (innovative), and test specific hypotheses about relationships between individuals and their environments (scientific) (Van der Ryn & Silverstein, 1967).

Hadjri and Crozier emphasize that an organization's largest investment is in human capital, their employees and others who may use their facility. By assessing a building with regard to its occupants and their experiences and behavior, they "provide the means to make their most valuable asset, their staff, work more productively" (Hadjri & Crozier, 2009, p. 27). Van der Ryn and Silverstein acknowledged the difficulty in connecting the value of a functional space to an organization's financial expenditures, but also its responsibility for supporting occupants by providing a physical environment that is appropriate, efficient, and effective.

> *The costs of environmental analysis should relate to improving the effectiveness of the institutional program. The consequences of environments that don't work well for people are often difficult to measure directly in relation to the institutional balance sheet. Often it is the user and his community, rather than the institution, who pay the psychic and social consequences of ill-fitting environments. (Van der Ryn & Silverstein, 1967, p. 82)*

Occupancy evaluations are meant to provide valuable information to a wide range of key players. Preiser summarizes this intention and value across the industry in an argument for broader adoption:

> *Both successes and failures in building performance are considered in POEs. The purpose of POE is to seek facts and not faults and to share the findings of POEs through dissemination and the creation of information clearinghouses. The entire building industry can benefit from this information through improved building quality and better value for the dollar. (Preiser et al., 1988, pp. x-xi).*

Challenges and Barriers to Occupancy Evaluation

Despite the numerous benefits and the momentum that seemed to be growing around POEs in the 1980s, occupancy evaluations are still only conducted on a small fraction of building projects. Zimmerman and Martin (2001) suggest that the barriers to widespread adoption include "fragmented incentives and benefits within the procurement and operation processes, lack of agreed and reliable indicators, potential liability for owners, exclusion from current delivery expectations [and] exclusion from professional curricula" (p. 168). They explain that the traditional design delivery process does not recognize continual improvement or involvement of designers beyond the design phases (Zimmerman & Martin, 2001).

Evaluations are not seen as part of architects' "normal services," and it is difficult to get clients to pay for them (Cooper, 2001). In addition, while there is great value in understanding the shortcomings of a facility, especially considering the tradeoffs that occur to meet budget and other constraints, there is uncertainty around problems that may be unearthed and the responsibility or liability to address these issues (Bordass et al., 2001). Due to the lack of a clear and single beneficiary in the evaluation process, with potential advantages and risks for clients and designers alike, there is reluctance on both sides to

https://doi.org/10.52202/067871-0003

fund the effort (Hadjri & Crozier, 2009).

Spectrum of Methods and Data

In addition to traditionally limiting occupancy evaluations to the post-occupancy phase, to many design practitioners, POE has become synonymous with surveying building occupants. However, even early texts on occupancy evaluations recommended a comprehensive and multi-faceted approach to studying building use. One of the first POEs, completed by Van der Ryn and Silverstein at the University of California, Berkeley in 1967, focused on college dorm residents and utilized a multi-method approach, through informal and structured observations and interviews, literature searches, user questionnaires, and self-reported activity logs, applied in three stages (Preiser et al., 1988).

Occupied buildings are a complex system of various components, from structural stability and technical performance and efficiency to user encounters with layout, space, and furniture; individual behaviors; and interactions with other occupants. It cannot be expected that a single method could capture all that is happening with and in a building, nor should we anticipate that such an approach is possible or even desirable.

Evaluations might range from the very simple, such as a facility walk through to assess the building use and condition in an unstructured and high-level manner, to a study of many facets of occupant use and behavior through a triangulated multi-method approach, engaging various types of users through several qualitative and quantitative forms of data collection (see Figure 2). The key drivers that determine where an evaluation falls on the spectrum of methods are the project's level of research integration, the degree to which a study aligns with the project objectives, and the level of commitment from the project team and client.

Figure 2

Spectrum of Occupancy Evaluation

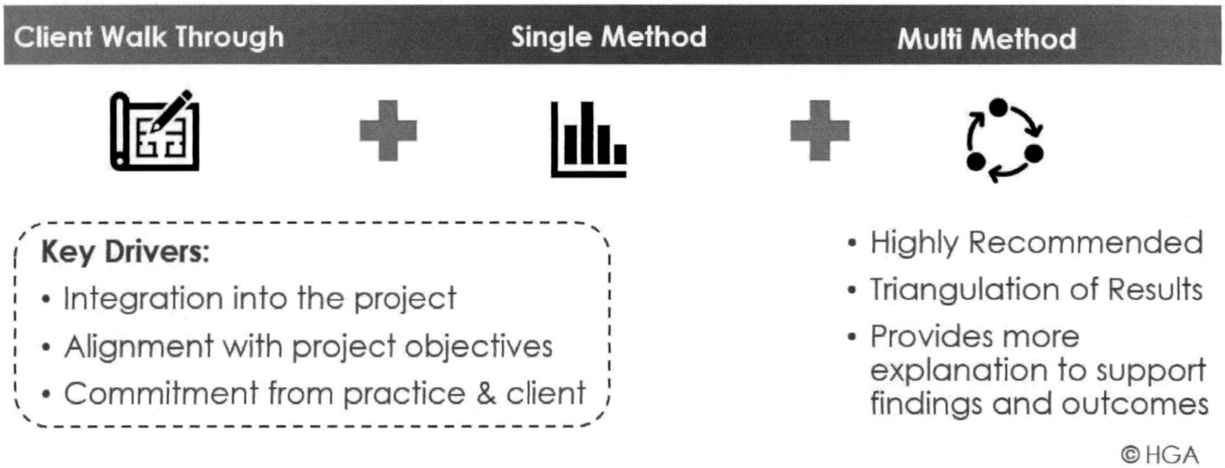

Depending on the level of evaluation rigor and investment, various methods, which generate different types of data, may be utilized to collect and analyze data related to occupant experience and outcomes (see Figure 3). Subjective methods engage occupants in an active manner, providing opportunities for direct feedback and self-reporting. Surveying, such as through questionnaires, follows a highly structured and prescriptive form and is often completed independently by participants. These tools may utilize an online or paper format for data collection and can accommodate a large number of participants asynchronously. Facilitated discussion methods, like focus groups or interviews, typically engage one or a small group of occupants in a more informal or semi-structured manner. Data are often qualitative in

 https://doi.org/10.52202/067871-0003

nature and provide depth and meaning to quantitative feedback received by surveying.

Objective methods typically capture data in present time as they occur. Monitoring/surveillance methods document occupant behavior, through shadowing people as they move throughout the facility or observing those who use a specific space. Occupant behavior data may also be collected with tracking devices or sensors. Since monitoring/surveillance methods document what people do, rather than what they say they do, these techniques assume an important place in environmental research. Physical readings can also be collected in real-time through specially calibrated instruments. These measurements are often tied to aspects of the ambient environment, such as acoustics, thermal comfort, or lighting.

Retrospective data collection includes a review of prior evidence from peer-reviewed literature or other sources, such as industry standards or benchmarking databases. Outcome metrics can also be received from the project organization. Organizational records often relate to occupant outcomes already monitored for other purposes, like hospital length of stay or student grades. Building performance can be tracked through collected energy metrics and utility expenses.

Although *prospective* methods are typically used in conjunction with data collected through other methods already mentioned, the category of simulation modeling is unique in that it can also serve as a method of generating data, based on a set of established or hypothetical parameters. Simulation may be used to test for potential human behavior and interaction through space syntax (discussed in the Saif Haq paper in these proceedings) or building performance scenarios, such as heat mapping of daylight exposure. These methods provide a valuable means for understanding spaces in an exploratory or iterative manner and the potential impact of various design scenarios on occupants without a physical building. Such approaches offer a big advantage in the design research field, where real-life experiments are very expensive and take many years to design and build.

Figure 3

Research Methods

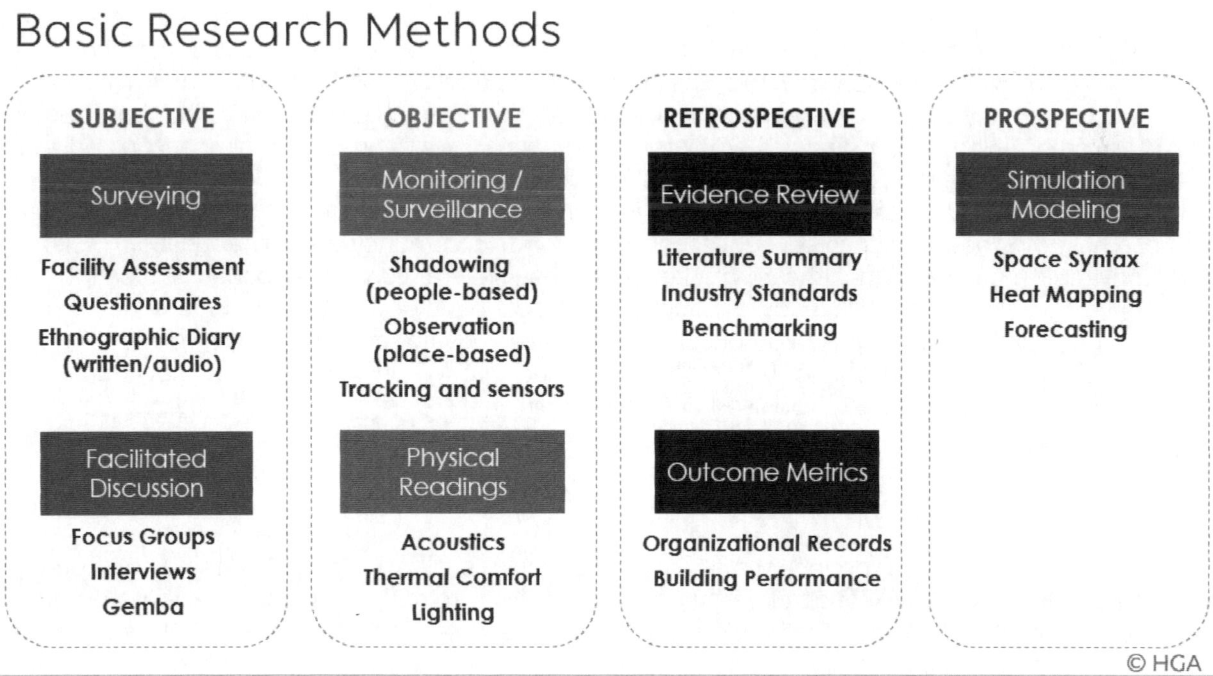

https://doi.org/10.52202/067871-0003

HGA's Strategic Occupancy Evaluation Approach

While significant challenges remain, the value of performing occupancy evaluations has become increasingly clear as more facility owners are realizing how to apply insights gained from continuous improvement activities in current environments and to future capital projects. Design teams are also realizing the value of basing design decisions on available evidence, including insights gathered from past projects. As a result, formal occupancy evaluation programs are taking shape at firms and organizations across the country. Job opportunities focused on the management, development, and curation of occupancy evaluation programs are becoming more common, heightening the demand for skilled researchers and practitioners.

With these trends in mind, HGA is expanding upon this definition of occupancy evaluation; viewing through the lens of a complete project timeline, we create opportunities for learning, immediate and ongoing operational improvement, and measurable comparisons of performance and functionality through a holistic, process-oriented lens. At HGA, occupancy evaluation embraces several key components:

1. Employs a rigorous and systematic research approach
2. Assesses facilities in use at any time point
3. Understands the occupant experience and perspective
4. Relates environmental elements to design intent, project goals, and operational needs
5. Identifies factors of success and opportunities for continuous improvement and learning

Through structured feedback from occupants, it is possible to connect the influence of environmental elements and design strategies to key outcomes and hypotheses, to understand if design intentions and project goals are met and operational performance is achieved. Evaluations are meant to uncover both successes and opportunities for improvement at any project phase—preceding a new project to capture the needs of occupants in an anticipated new space, following construction and occupancy as evidence and assessment of achieved goals, or as part of a planning process to understand the current state and drive future strategic decisions. In addition, a phased occupancy evaluation has been explored as a means to gather data at different periods following occupancy of a building to address operational and workflow inefficiencies recognized through data. Once recommended changes are operationalized, another 'phase' of data collection monitors and evaluates the updated state, creating a continuous improvement occupancy evaluation approach (Freihoefer & Zborowsky, 2017). Overall, a standardized and systematic methodology applied across projects and sectors allows for in-depth comparison and learning, and the creation of a robust database of baseline occupancy metrics.

Similar to many architecture and engineering firms, HGA is composed of many different market sectors. These include healthcare, public/corporate, arts/community/higher education, science/technology, and more. HGA's healthcare market sector has several years of experience creating a research-driven practice and has already conducted post-occupancy evaluations on many hospital and clinic projects. Validated tools for data collection have been developed, refined, and piloted. Furthermore, a protocol for deploying the tools has been approved through an Institutional Review Board (IRB). The tools are sufficiently versatile to be used during any point in the design process. Gathering data is especially useful during pre-design to then apply to design solutions. The successful development of standard tools that can scale across various healthcare project types and phases of design (e.g., pre-design, pre-occupancy, and post-occupancy) with ease and swiftness is a key component of a robust occupancy evaluation program.

Another important component of a scalable occupancy evaluation program is finding opportunities to adapt tools across market sectors. Many markets sectors are researching similar design problems. These include collaboration and mentorship, workplace strategy, efficiency and productivity, security and safety, and wellness and wellbeing. Therefore, many of the standardized tools and protocols created to evaluate occupancy within one physical environment can be adapted across other market sectors and projects.

As a strategic priority at HGA, development of an occupancy evaluation program has been many years in the making. A key success in achieving buy-in from our internal stakeholders (e.g., executive lead-

 https://doi.org/10.52202/067871-0003

ership, market-sector leaders, design teams, etc.) has been to align the program with strategic priorities of the market. Because of the number of projects HGA designs every year, it is impossible to conduct an occupancy evaluation on each one. Instead, dedicated advisory groups for each market sector help prioritize which projects to evaluate. Advisory groups are interdisciplinary groups of thought-leaders, made up of 8 to 12 colleagues within each market sector, composed of the sector leader, principal architects, interior designers, engineers, market strategists, business developers, and others. Each advisory group has a clear understanding of the pulse of the industry client needs, knowledge gaps, and trends in the market. HGA's research team regularly meets with advisory groups to identify potential occupancy evaluation projects and to align project types and research inquiry with the market sectors' strategic goals, which may include areas or topics each market sector wishes to target, know more about, or become more competitive in.

Each market sector earmarks funding to fully or partially sponsor occupancy evaluations every fiscal year. Oftentimes clients are willing to fund part or all of an evaluation study because they garner as much valuable insight from the evaluation as do the design firms. Occupancy evaluations have become a joint effort to examine occupants' perceptions and behaviors along with overall building performance. It is a win-win scenario for all parties involved to further advance design and the overall industry.

Conclusion

Through the establishment of clear goals and objectives, a building design is intended to provide effective qualities at an appropriate value while minimizing problems. Thus, the role of an evaluation is to compare actual performance with explicitly stated criteria of success carefully developed from the highest level of understanding of the occupants and their needs (Van der Ryn & Silverstein, 1967).

Research and evaluation provide the highest level of value when incorporated throughout the design process and in a cyclical manner, bringing learning from one project to inform the next, and providing relevant information at various stages. Ultimately, a successful occupancy evaluation program should offer a beneficial service to project teams, management, clients and prospective clients, industry colleagues, even competitors, through practical and applicable knowledge of how buildings are used and ways to incrementally improve design for the wellbeing and productivity of those who use them. From this foundation comes opportunities to develop compelling stories connecting design impact to evidence through a comprehensive database of occupant feedback using a consistent process across all market sectors and project types (Bordass & Leaman, 2005).

In addition, Bordass and Leaman (2005) call for standard measures across the industry and open-source benchmarks accessible through managed structured databases. Although there have been discussions in the design research field about the potential to achieve this level of access and synergy across the industry, it sometimes seems difficult to imagine its existence. However, HGA's occupancy evaluation program is an important step toward this ideal.

References

Bordass, B., Leaman, A., & Ruyssevelt, P. (2001). Assessing building performance in use 5: Conclusions and implications. *Building Research & Information, 29*(2), 144-157. https://doi.org/10.1080/09613210010008054

Bordass, B., & Leaman, A. (2005). Making feedback and post-occupancy evaluation routine 1: A portfolio of feedback techniques. *Building Research & Information, 33*(4), 347-352. https://doi.org/10.1080/09613210500162016

Cooper, I. (2001). Post-occupancy evaluation-where are you?. *Building Research & Information, 29*(2), 158-163. https://doi.org/10.1080/09613210010016820

Freihoefer, K., & Zborowsky, T. (2017). Making the case for practice-based research and the imperative role of design practitioners. *Health Environments Research & Design Journal*, 10(3), 66–82. https://doi.org10.1177/1937586716674470

Hadjri, K., & Crozier, C. (2009). Post-occupancy evaluation: purpose, benefits and barriers. *Facilities, 27*(1/2), 21–33. https://doi.org/10.1108/02632770910923063

Parshall, S. (1989). A hospital evaluation: The problem-seeking method. *In W. F. E. Preiser (Ed.), Building evaluation (pp. 207-220).* Springer.

Preiser, W. F. E., Rabinowitz, H. Z., & White, T. (1988). *Post-occupancy evaluation*. Van Nostrand Reinhold.

Van der Ryn, S., & Silverstein, M. (1967). *Dorms at Berkeley: An environmental analysis*. University of California, Berkeley Center for Planning and Development Research.

Zimmerman, A., & Martin, M. (2001). Post-occupancy evaluation: Benefits and barriers. *Building Research & Information, 29*(2), 168-174. https://doi.org/10.1080/09613210010016857

 https://doi.org/10.52202/067871-0003

Five Essential Decisions for Clarity of Space Syntax Methodology

Saif Haq, *BArch, MS, PhD (College of Architecture, Texas Tech University)*

ABSTRACT

Human understanding of environments often develops through asynchronous experiences. Individuals are either peripatetic, learning about the 'relations' between environmental units as they move about, or, if stationary, experience the effects of diurnal and seasonal changes. Either way, what they 'see' is always changing. The environmental learning of mobile humans is often called a 'cognitive map', and in turn is related to moving behavior and navigational decision making.

Modelling the effects of continuous changes of location with sufficient rigor for quantitative analyses has been challenging. The theory and methodology of Space Syntax offers precise ways of measuring complex environmental properties that arise from connections of each space to all other spaces contained in the same layout – sometimes directly, sometimes through other intermediate spaces. These analyses indicate a visual 'structure' perhaps comparable to that of a moving human's experience. As such, they can be understood as synchronous models of asynchronous experiences. Increased computational powers now allow finer tuned layout investigations by considering topological, metric, and geometric relationships, defining different kinds of unit spaces, and choosing the radii of analyses. However, since moving humans may or may not travel to or see all spaces in a layout, Syntax analyses might be considered 'idealized' because all spaces in a plan drawing and connections to all other spaces are the base of analysis.

This presentation will couple foundational concepts of Space Syntax with advanced analytical techniques. After a brief review of research that used these variables to investigate human and medical outcomes, it will conclude with a discussion of its drawbacks and opportunities. With respect to the agenda of this conference, the case studies and discussions will focus on healthcare and public health outcomes.

INTRODUCTION

Changes in our location bring attention to how different spaces are connected to one another. Properties of spaces based on such connections could be theoretical representations of our moving experiences. In this scenario, Space Syntax offers precise ways of understanding and measuring complex spatial properties that arise from connections that each space has to all other spaces contained in the same plan – sometimes directly, sometimes through other in-between spaces. Syntax analyses depict a visual 'structure', and this is perhaps comparable to the experience of a moving human. In other words, a Space Syntax model can be understood as a synchronous depiction of asynchronous experiences. However, since moving humans may or may not travel to or experience all the spaces in a layout, Syntax models could be considered an 'idealized' experience structure (see Figure 1).

Currently, increased computational powers allow finer-tuned layout modeling by considering topological, metric, and geometric relationships, defining different kinds of unit spaces, and choosing differ-

https://doi.org/10.52202/067871-0004

ent radii for analyses. With the possibilities of analysis increasing, five important decisions prior to selecting the Space Syntax methodology are essential. This paper describes those and emphasizes that this clarity is crucial at the onset of any study that uses Space Syntax.

Theoretical Concepts Underpinning Space Syntax

At the beginning of any discussion on Space Syntax, it is prudent to shed some light on some of its fundamental concepts that remain significant today. Most important is the idea that spatial connections are understood and identified based on the experiences of the moving human person inside an environment. From this starting assumption, Space Syntax then proceeded to compute quantitative values of unit spaces to mimic aggregates of the moving humans (Hillier, 1970, 1973, 1977; Hillier & Leaman, 1974; B. Hillier et al., 1972). Two important variables were named 'Integration' and 'Choice'. However, currently the more descriptive terms 'Closeness' and 'Betweenness,' respectively, are used as they are indicative of the professed relationship between spatial attributes and human experiences.

The second concept is a 'non-discursive' grasp of space (Hillier, 2007, chapter 3). This concept can be understood when we think of giving directions to strangers. Typical descriptions in such a condition might be "keep going straight and when you reach the intersection take a right", or "take a left at the junction". These examples illustrate that the concept of connections (between spaces) is at the root of our spatial understanding, even before we consider distances and/or shapes. Indeed, to first receive directions such as "go 536.9 yards, turn 96.3 degrees" would be uncommon, and perhaps disturbing, to say the least. Space Syntax is built on this fundamental spatial property – i.e., connections between spaces. Indeed, Hiller has stated "Relations, it seems, are what we think with, rather than what we think of" (Hillier & Hanson, 1984, p. 02).

The third aspect, related to the previous ones is that built spaces are permeable and hence continuous –apparently infinitely. A person can go out of the bedroom to the living room, to the front yard, to the street, to the grocery store, to the friend's house, etc., on and on. Because space is continuous, the idea of immediately connected spaces (called Level 1 connections) and spaces connected through one or more spaces (Level 2, 3 4, 5, etc. connections) can be comprehended. The idea of continuity makes even more sense when this experience is empirically investigated through the lens of spatial cognition, something that began at least two decades after Syntax was introduced (Dalton, 2003; Haq, 1999; Penn, 2001).

Finally, also from a human-centered understanding, ideas of synchronic and diachronic experiences are relevant. Considered from the perspective of Ferdinand de Saussure (2006), logical and psychological connections between coexisting items of a system, i.e., what we see 'at once' (passive observer) is synchronous learning, and "connexions (sic) between sequences of items not perceived by the same collective consciousness" (p. 06), i.e., what we experience over time is diachronic. Space Syntax calculates variables of each space by considering relations from each space in the plan to all other spaces in it, -- thus it is an idealized aggregate. In other words, if a person were moving from all spaces to all other spaces, then their experience would be captured by Syntax values. Thus, the Syntax representation of variables is diachronic information presented in a synchronic form, i.e., simultaneously in a plan diagram (see Figure 1).

 https://doi.org/10.52202/067871-0004

Figure 1

A Plan and a Space Syntax Model

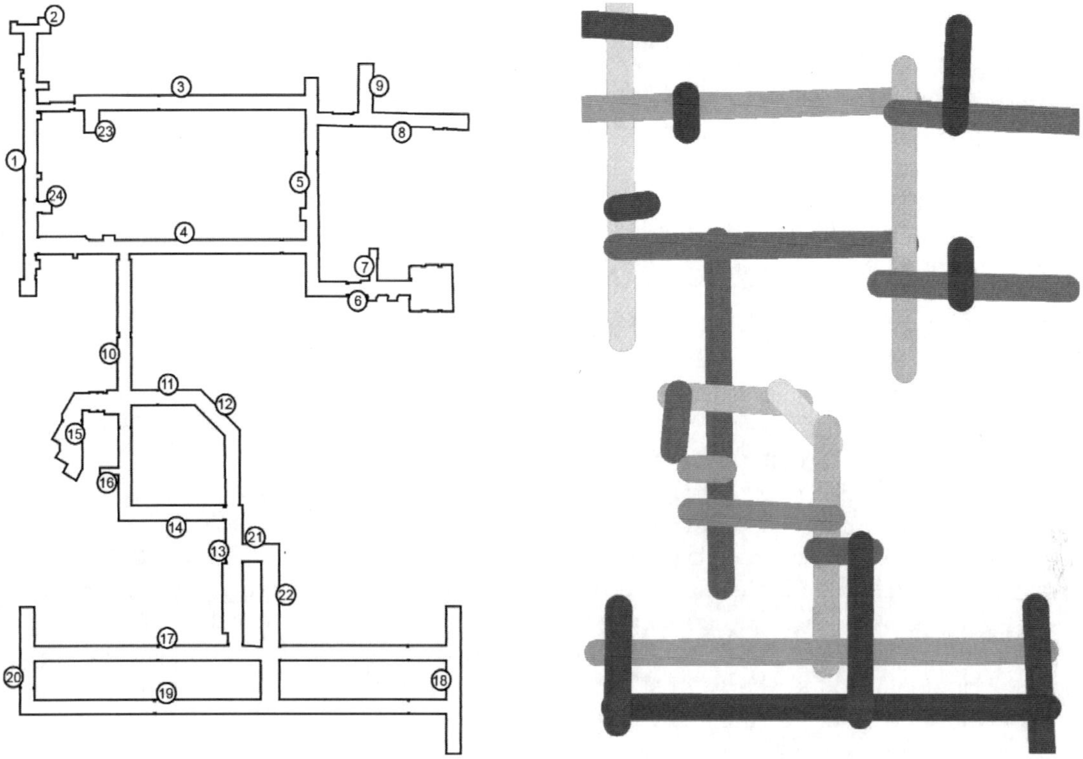

Note: Corridors are unit spaces and topological connections to the entire plan are considered. Colors represent Closeness(n) values with higher to lower values depicted from red to blue colors. Source: Author

Introduction to Methodology

Space Syntax is both a theory and a methodology. It focuses on the relations between spaces as seen in plan drawings of various scales, quantifies each space based on these relations, and suggests that they have associations with numerous human aspects. Space Syntax began in the 1970's by considering topological relationships between unit spaces (in a plan) and hypothesizing a connection with social patterns -- thus the title of its first book was The Social Logic of Space (Hillier & Hanson, 1984). With time, some operational decisions made at the beginning stages were reconceptualized and elaborated, sharper ideas about spatial relationships and calculation methods of spatial variables were put forward, and more robust connections to human functions were postulated. In the process, powerful computer software to analyze large plans, and to include the newer conceptualizations of human attributes were developed.

Access to these digital tools has made analysis of spatial connections easy, and worldwide recognition of Space Syntax has spurred additional empirical research to investigate associations to other kinds of aggregate human-space relationships. These investigations have led to quite a few useful results. For example, Space Syntax variables were found to be positively correlated to pedestrian movement patterns in cities (Hillier et al., 1987; Peponis et al., 1989), traffic flows (Penn et al., 1998), land values (Kubat, 2009), wayfinding use of corridors inside buildings (Haq & Zimring, 2001; Peponis et al., 1990), spatial cognition (Haq, 1999; Hölscher et al., 2012; Penn, 2001), depression symptoms in homes (Chambers et al., 2018), etc. In hospital settings Space Syntax was related to nurse and physician positioning (Lu & Zimring, 2012), nurse entries to rooms (Hendrich et al., 2009), bed preferences by patients (Alalouch et al., 2009), and mortality rates (Ossmann, 2016), among others.

https://doi.org/10.52202/067871-0004

Advances in Space Syntax concepts and developments of the software took place concurrently with one influencing the other. For example, new ways to define unit spaces were developed and new relationship patterns between them identified. These were programmed into the software allowing newer kinds of layout analysis, and novel hypotheses relating them to unexplored aggregates of human attributes were tested.

As more and more research is reported, a clarity of Space Syntax methodological decisions is needed both to support the research questions posed, and to make sure that the selected variables and analyses are appropriate. Since Space Syntax has become much more fine grained and precise, it is no longer enough to simply provide very general statements such as "Space Syntax research suggests.....". This paper identifies and explains five decisions about Space Syntax factors that must be made in the beginning of any spatial analysis and explained thoroughly.

Foundational Ideas Behind Space Syntax Methodology

As previously indicated, the original idea of Space Syntax methodology was based on connections that each space has with all other spaces in a plan, either directly to adjacent spaces, or through a set of other spaces. Based on this notion of connections, and considering all spaces as both origins and destinations, Space Syntax calculated numerical values of each space.

The diagrams that follow are offered as a simplified explanation of Space Syntax initial ideas. In the plan shown in Figure 2 (right side), Corridors 1, 4 and 10 are directly connected to a set of adjacent corridors, and these connections are shown in red, blue, and green curved lines respectively. The connection type considered in the original Space Syntax methodology was topology. In other words, Syntax only considered whether a space was connected to another one or not. Size and shapes of the spaces were inconsequential. Thus, each space could be reduced to a dot with lines representing connections to other spaces – bringing forth the characteristics of a graph. The immediate connections of Corridors 1, 4 and 10 in graph-form are shown on the left of Figure 2.

Figure 2

Connections Between Different Spaces

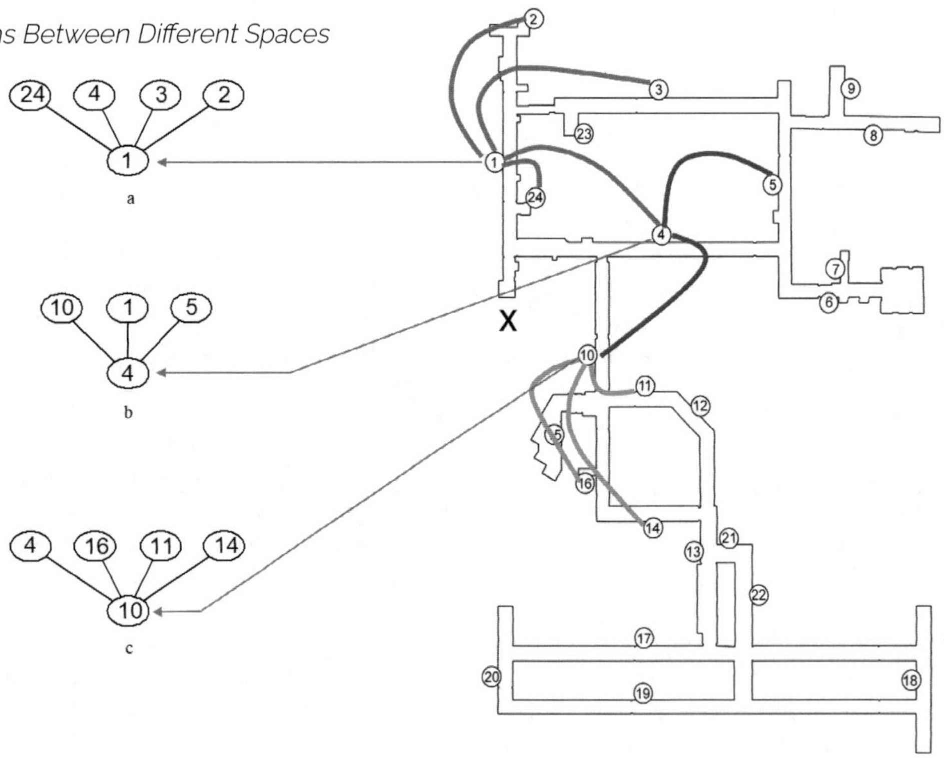

Note: Connections between different spaces are identified (right) and shown in graph form (left).
Source: Author

 https://doi.org/10.52202/067871-0004

This idea of connections can further be extended to indicate secondary, tertiary, and more distant connections. In other words, a space can be connected to immediately adjacent ones, and, through some spaces to others further away, and so on. To illustrate, using Figures 2c and 3b, Corridor 10 is connected directly to Corridors 4, 6, 11 and 14 (Figure 2). Further, 4 is connected to 1 and 5, 1 is connected to 2, 3, 24, and so on (see Figure 3b). Thus, the notions of direct connections, secondary connections, tertiary connections, etc. are understood.

Figure 3

Connections of Corridors

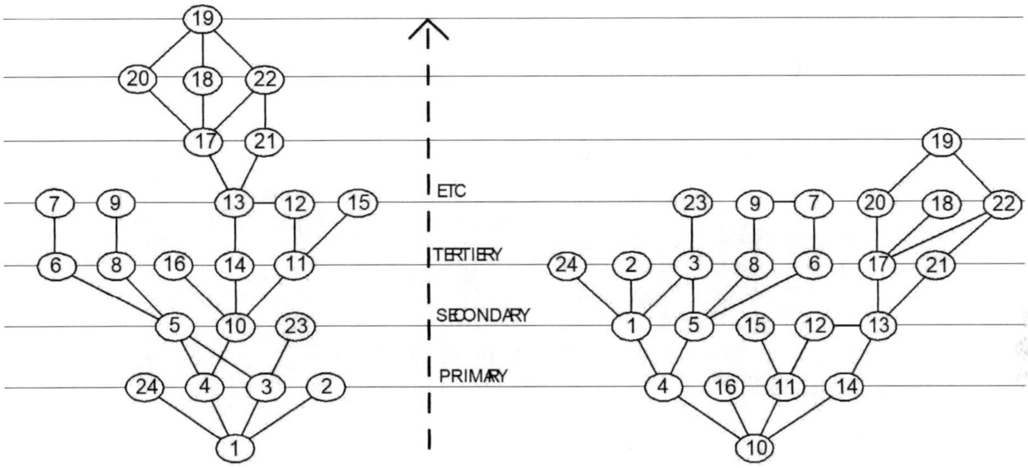

Note: Connections of corridors 1 and 10 to all spaces (in the plan in Figures 1 and 2) are shown as a graph. 3a shows connections from corridor 1 and 3b shows them from corridor 10. Source: Author

Space Syntax calculations look at each space's connections to all other spaces in the plan, and after comparing those to an idealized plan, generates two kinds of numerical values for each space – Integration and Choice[1]. The Integration value indicates how close an origin space is to all other spaces in the plan, and Choice measures how often a space lies on the shortest paths (connections) between any pair of spaces (Hillier & Hanson, 1984). Recently, these have been replaced with the newer terms 'Closeness' and 'Betweenness' for ease of understanding, and this approach is also used in this paper.

Space Syntax: Five Important Considerations

As mentioned earlier, Space Syntax theoretical concepts have been developed, and methods have become more sophisticated. Fortunately, free software is easy to download and use. As such, plans can be effortlessly modelled with different assumptions. However, doing a computer analysis without a thorough understanding of the concepts is antithetical to developing valuable research questions and selecting methods. This paper proposes five aspects that should be carefully considered, and decisions clearly explained in all research that uses Space Syntax. Each factor is described with appropriate images in the text that follows.

1. Unit Spaces

In that narrative presented previously, spaces have been addressed simply as unit spaces (Corridors were the unit spaces in Figure 2). While a space is indeed a unit in Space Syntax analysis, there needs to be a clear definition about what a unit space (for analysis) is. The first task, therefore, is to describe what these unit spaces will be, and to postulate what it means for the research being undertaken.

When it began, Space Syntax considered Axial Lines as unit spaces (Hillier & Hanson, 1984). Such

[1] Other variables are also calculated, but Integration and Choice are the two mostly used.

https://doi.org/10.52202/067871-0004

lines were described through the concept of Convex Spaces. The idea about convex space came from human awareness, and was defined as that space where all points in its perimeter would be visible from all the points within it (Hillier & Hanson, 1984). The task of the researcher would be to reduce the plan to a set of 'largest' convex spaces (see Figure 4b). It is important to understand that this process had a subjective component when spaces of complex shapes were considered because they had to be manually 'broken up' into convex pieces. Then, all the convex spaces in the plan had to be connected using the fewest and the longest lines possible (Hillier, 1999). These were called Axial Lines and were then considered the unit spaces in Space Syntax analysis (see Figure 4c). Over time, axial lines were directly drawn to represent linear spaces such as building corridors and city streets (without including the notion of convex spaces).

Figure 4

Open Space System to Convex Space and to Axial Lines

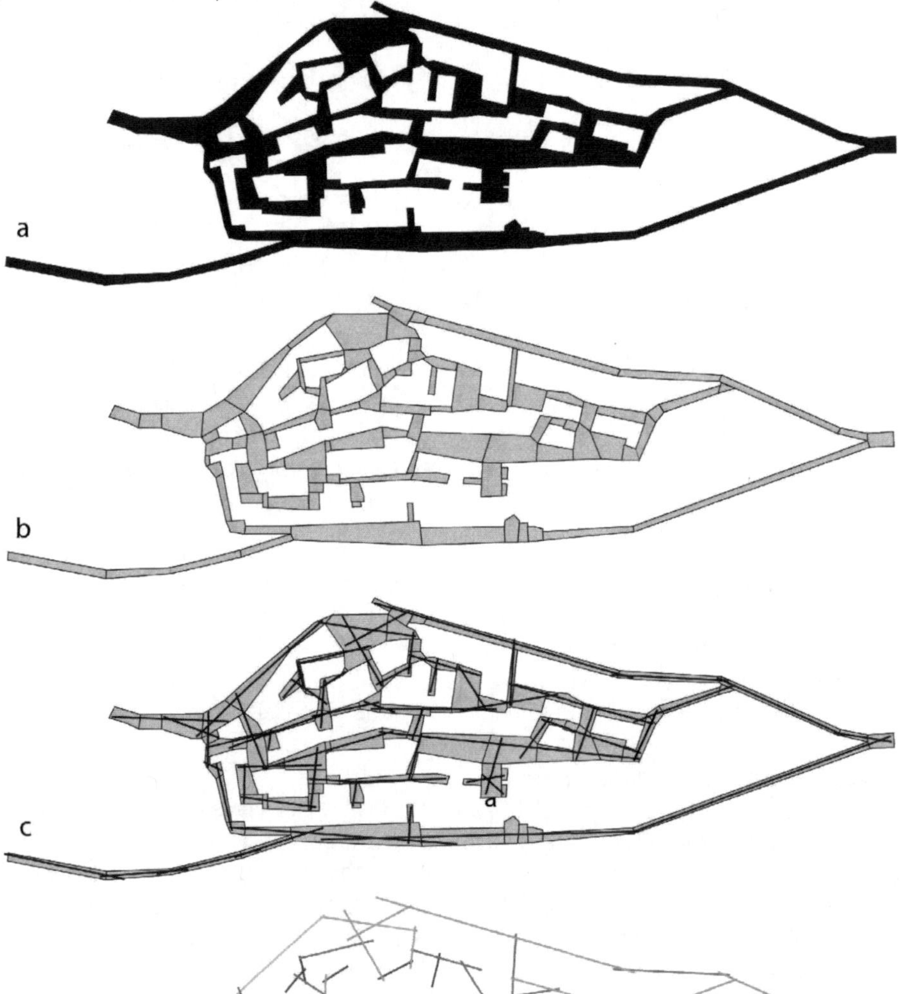

Note: (a) Open space system of a small town (streets and squares) shown in black (b) Open space system divided into a set of convex spaces, (c) Longest and fewest (axial) lines drawn to connect all the convex spaces, (d) Closeness values of axial lines. Source: Author

 https://doi.org/10.52202/067871-0004

New versions of Space Syntax software can now generate axial lines automatically. The software can produce all possible axial lines as depicted in Figure 5a. Since the software produces too many lines for meaningful relationship with real spaces, it can automatically reduce the number of lines too (Figure 5b). However, the possibility of manually drawing the lines is also available. Occasionally, this option might be used to match the analysis with the research question being asked (see Figure 5c).

The last kind of axial lines are called segmented lines, where the axial lines are segmented at their intersections to create another kind of axial map (see Figure 5d). In such a condition, the length of each line is reduced at their segments and the total number of lines in the plan increases. Please note that axial lines produced in any of the three methodologies can be segmented. This approach is especially useful in urban conditions, as shown in Figure 5e.

Figure 5

Examples of Different Axial Lines and Their Analysis

Figure 5a. Computer generated all possible axial lines and their topological analysis. (Closeness values) Source: Author

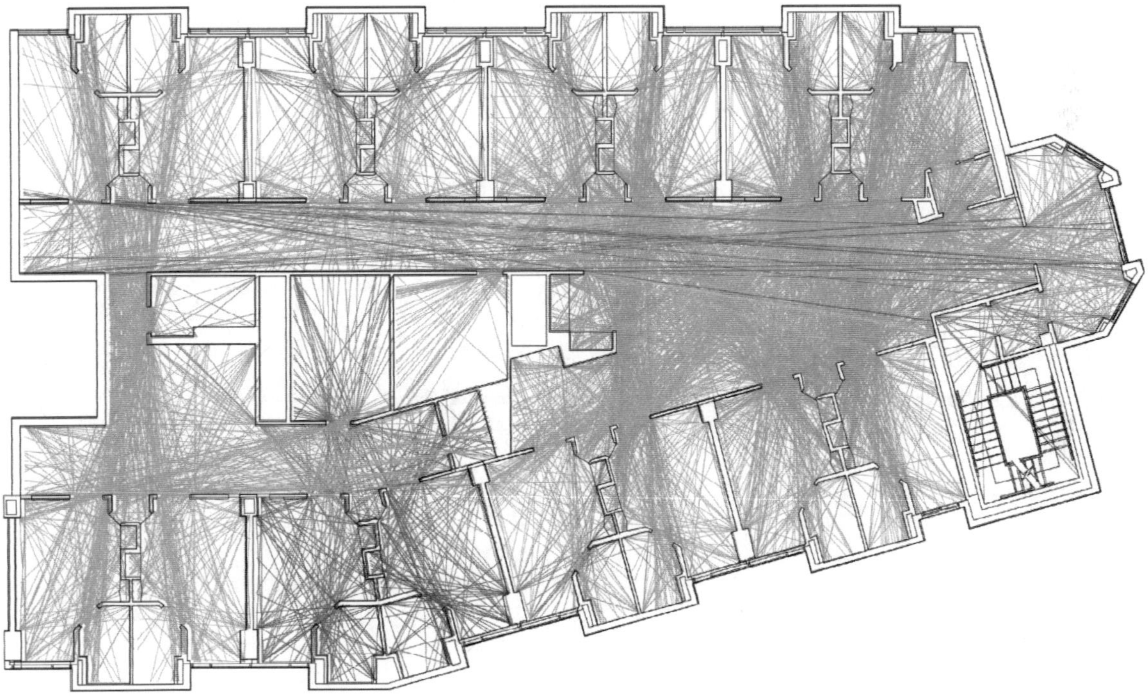

Figure 5b: Computer has automatically reduced the number of lines from those shown in 5a. (Closeness values) Source: Author

Figure 5c: Axial lines manually drawn with digital tools to correspond specifically to the research intention. (Closeness values) Source: Author

 https://doi.org/10.52202/067871-0004

Figure 5d. Axial lines are segmented at their intersections to create a segmented line map. (Betweenness values) Source: Author

Figure 5e. Streets of Dammam, Saudi Arabia, modelled as segmented lines showing Closeness values. Source: Alrashed (2021)

https://doi.org/10.52202/067871-0004

It may not be farfetched to say that the ubiquitous use of axial lines in the first few decades of Space Syntax software was due to two things: (1) most research undertaken was in urban areas where streets could be naturally (and quickly) represented as lines, and (2) the software could only deal with axial lines. The reader is reminded that the theory of Space Syntax was never limited to lines alone. Recent advances of computing now allow researchers to define their own unit space. In addition to axial lines, the following unit spaces have been used in research: rooms (Figure 6a) and units of a grid superimposed on a plan[2] (see Figure 6b). Additionally, researchers have the option of defining their own units and use the software for plan analysis. What is important is that the unit spaces are clearly defined, and the reasons explained with reference to the research question being posed.

Figure 6

Examples of Unit Space and Analysis

Figure 6a. Rooms of this house is used as unit spaces for analysis. (Closeness values) Source: Author

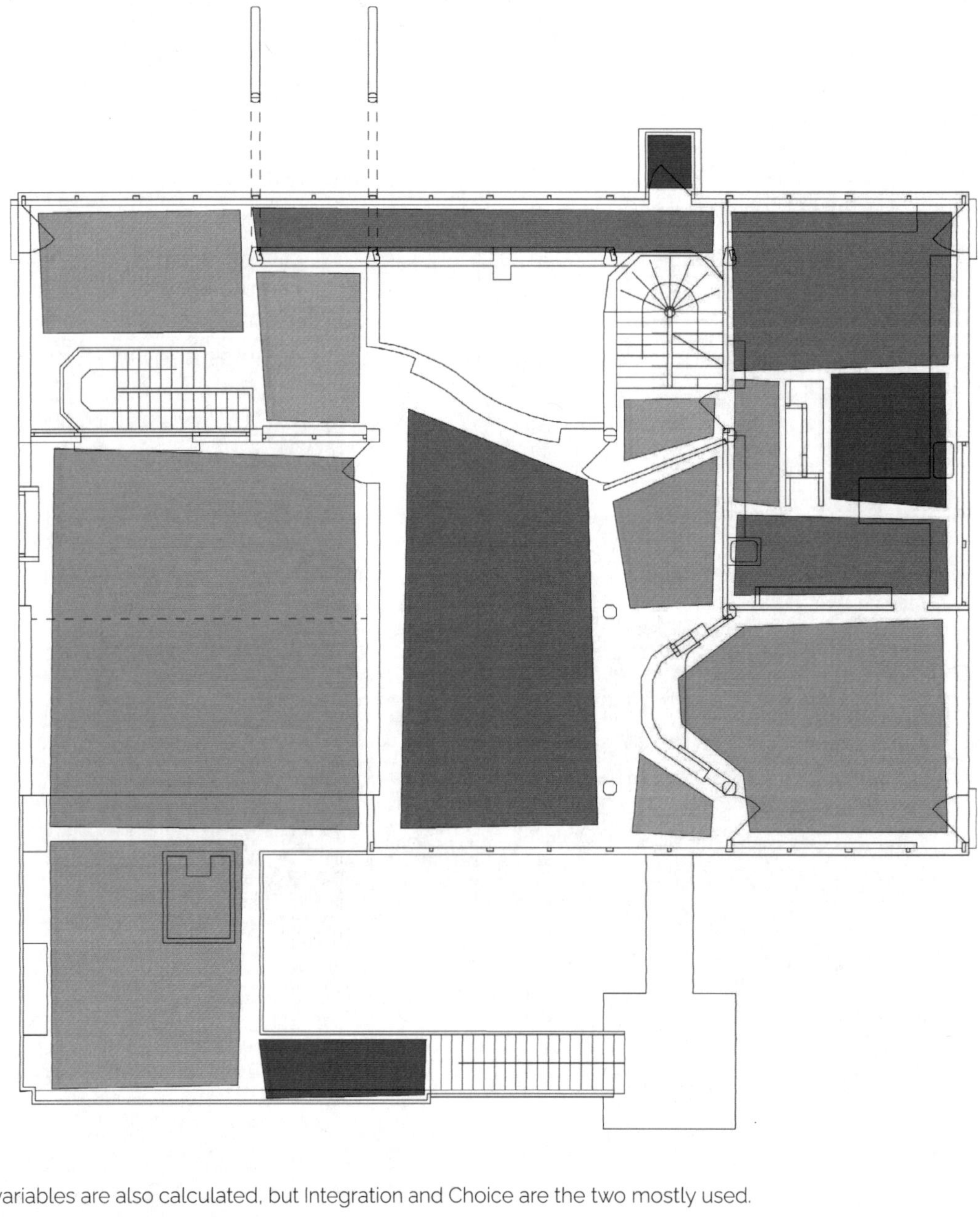

[2] Other variables are also calculated, but Integration and Choice are the two mostly used.

 https://doi.org/10.52202/067871-0004

Figure 6b: A grid of appropriate dimensions is laid over the plan and each 'tile' is used as a unit space in this analysis. Source: Author

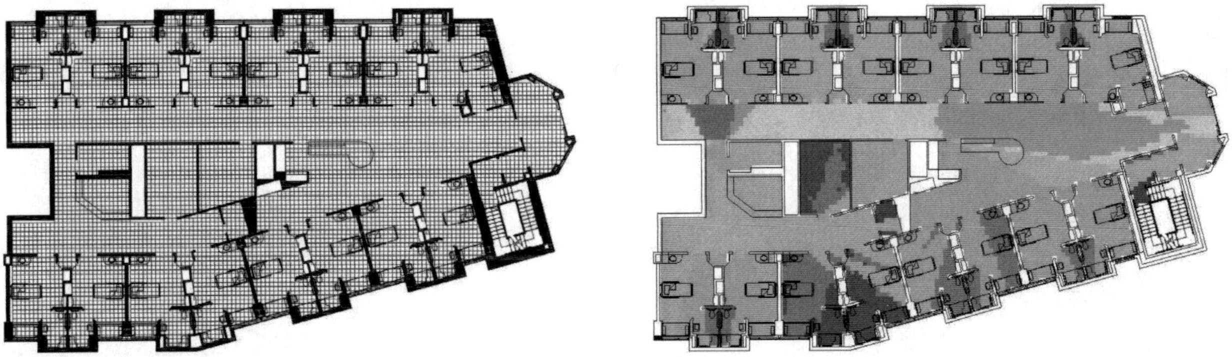

2. Spatial System

A spatial system is a set of researcher-defined unit spaces connected to form a layout or a plan. Layout is the idea of spaces configured together without a sense of enclosure or boundary, whereas a plan is a definite architectural drawing representing a building or an area. In other words, a plan has a boundary, albeit invisible, all around it. This boundary restricts the infinite continuity of spaces (described in Section 1 above) and gives an indication of where a moving person might stop, or a computer might stop counting. Computation by Space Syntax software is bounded within the plan diagram that the researcher selects. It is important to understand that depending on where the boundary is set, values of each space will vary (see Figure 7). Thus, the boundary definitions must be properly argued. In quantitative analysis such variations in values might seriously jeopardize the conclusions.

Definition regarding the boundary is a decision of the researcher, but it must be selected after careful deliberations. These deliberations should be clearly explained so that the Syntax analysis will be appropriately related to the research intentions.

Figure 7

Examples of Different Axial Lines and Their Analysis

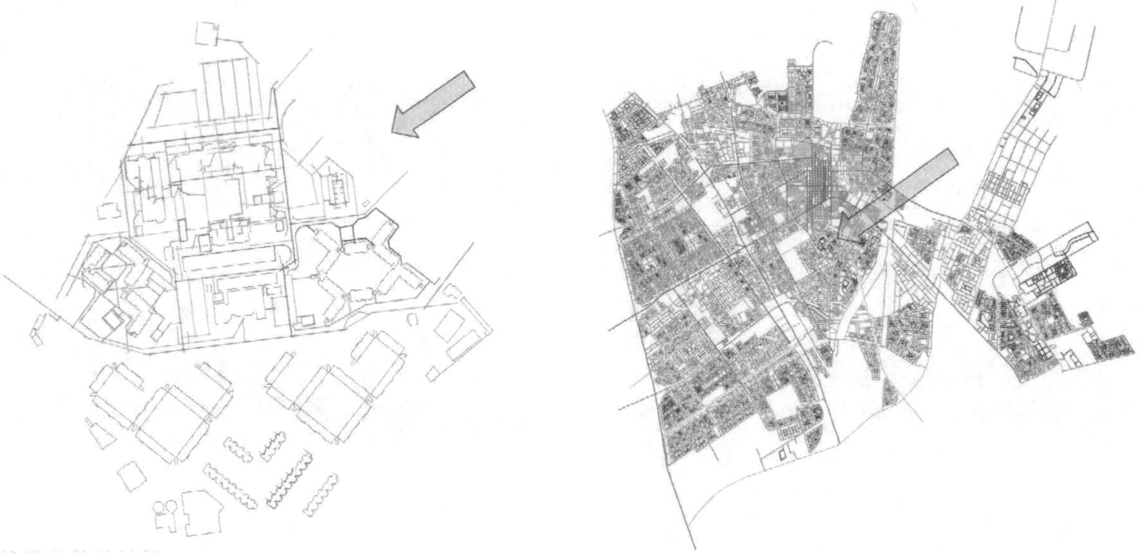

Note: Streets inside King Fahad Specialist Hospital is modelled on the left, and all streets in the city of Dammam including the hospital complex, is shown on the right. The Syntax value of any street inside the campus will have two values depending on which model is selected. Source: Alrashed (2021)

A related notion regarding the Spatial System is deciding what is to be included inside the boundary. In Figure 8 that follows, public corridors are considered in 8b, and all spaces in the plan are selected in 8c. Such decisions will also change the Space Syntax variables and have an influence on the empirical work being carried out.

Figure 8

Example of Plan and Axial Analysis

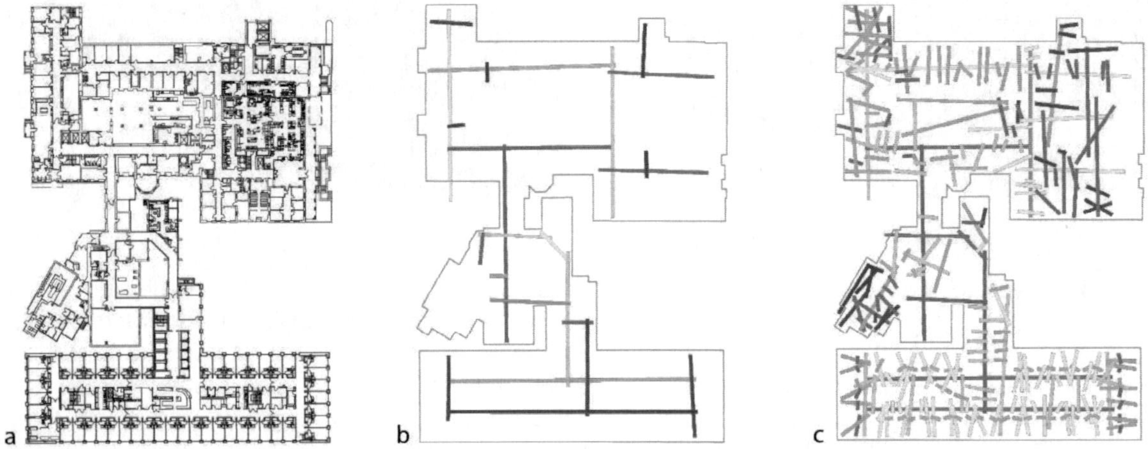

Note: Figure 8a, Plan of a hospital, 8b Axial analysis of public corridors and 8c Axial analysis of all spaces. Source: Author.

Finally, a third factor in the selection of the Spatial System comes from research intentions. For example, if visibility is being considered, then the furniture and interior glass partitions in the plan are ignored for modelling because humans can see over low furniture and see-through glass. On the other hand, when dealing with accessibility, then both furniture and glass walls are drawn because they impede where humans can walk to in the plan (see Figure 9).

Figure 9

Example of Accessibility and Visibility

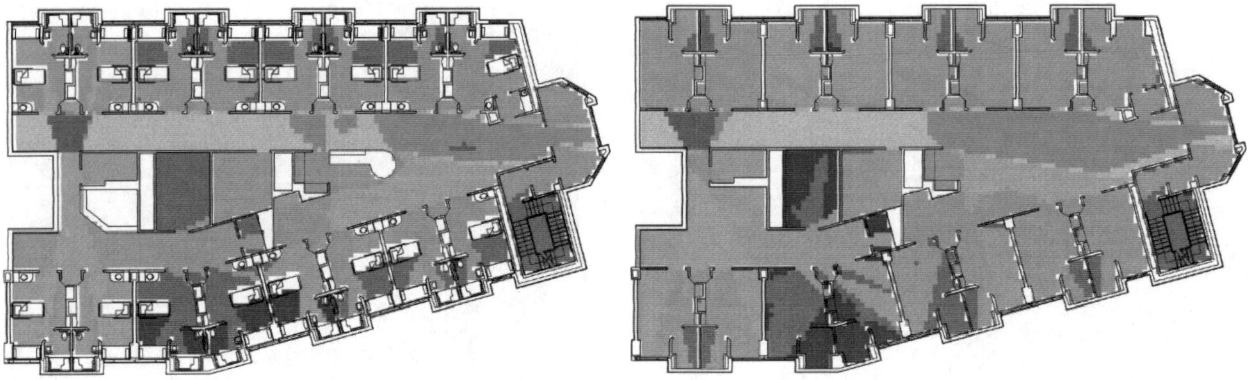

Note: Accessibility (left) and Visibility (right) models created by selecting appropriate spatial system. Source Author

 https://doi.org/10.52202/067871-0004

3. Concept of Distance Between Unit Spaces, or Connection Types

It has been stated many times that relationships between spaces are the basis of Space Syntax analysis. However, the kinds of relationships were not defined. This definition, and further clarification are the third important decision that must be taken at the beginning of any study that uses Space Syntax. Essentially, with the development of software, Space Syntax can now consider three kinds of relationships --topological, metric, and geometric.

The genesis of Space Syntax emphasized topological relationships. Its roots coincide with Syntax's early description of human movement behavior, as described earlier. Topological relationships simply note if a pair of spaces are connected to one another, or not. Taking this simple relationship between spaces, Space Syntax has flourished for more than three decades. At present, two other kinds of relationships can be computed – metric and geometric.

Metric relationship considers the distance between two pairs of spaces in the calculation of closeness or betweenness, and geometric relationship looks at the angles between them. These concepts can be understood with the help of the diagram in Figure 10, which represents a set of axial lines. The centers of two lines are marked as X and Y. Considering metric distance, Path B is shorter than Path A. However, in terms of angles, Path A is shorter. From a topological point of view both paths are the same as they require only two turns. These factors can also be modelled for entire layouts in newer versions of Space Syntax analysis (see Figure 11).

Figure 10

Three Concepts of Distance Between Spaces

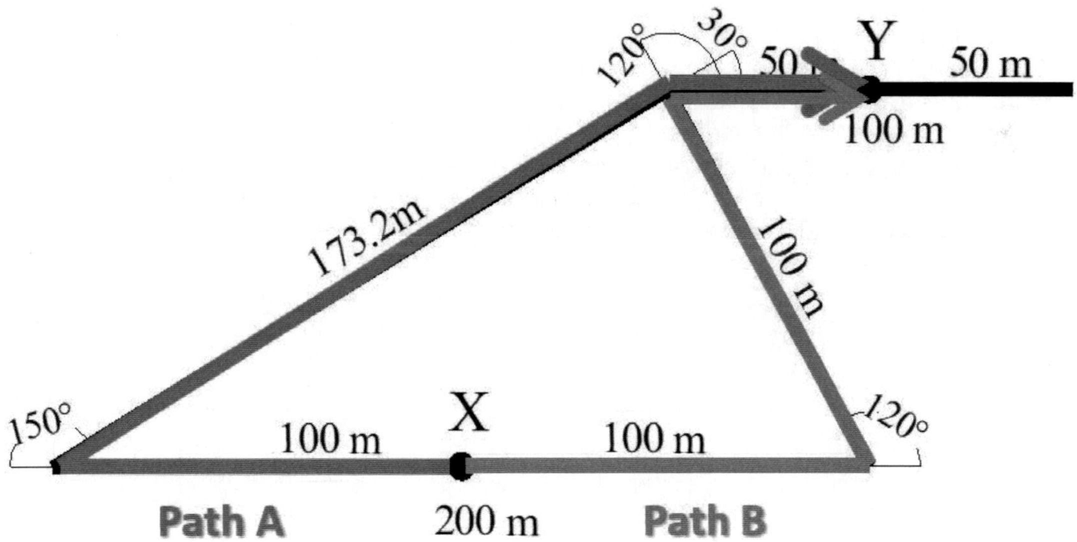

Note: Three concepts of distance between spaces: Topological, Metric, and Geometric. Source: Berhie (2016)

https://doi.org/10.52202/067871-0004

Figure 11

Topological, Angular and Metric Closeness

Note: Topological, Angular and Metric Closeness analysis of a small town using segmented lines as unit spaces. Source: Author

Decisions regarding the distance types between spaces, or connection types are important and must be clearly related to the research question.

4. Radius of Analysis

As explained previously, Space Syntax considers connections from all spaces to all other spaces in the selected spatial layout shown as a plan. However, analysis could be theoretically stopped at any topological, angular, or metric distance to compute another version of the desired variable (Closeness or Betweenness). This difference can be understood by comparing Figure 3 with Figure 12. In the latter, a line could be drawn at any distance from the origin. In this scenario, the topological distance was selected at '3' (see red line in Figure 12) then calculated values would be Closeness-3 or Betweenness-3, and different from the regular Closeness or Betweenness values (referred to Closeness-n or Betweenness-n.; see Figure 13). This 'radius of analysis' could be set anywhere the researcher decides. In many cases, a shorter 'radius of analysis' shows better relationships with human attributes (e.g., see Hillier (2005). It is important to clarify the 'radius of analysis' selected, and the reasons for such a choice need to be clearly explained in the beginning.

Figure 12

Radius of Analysis

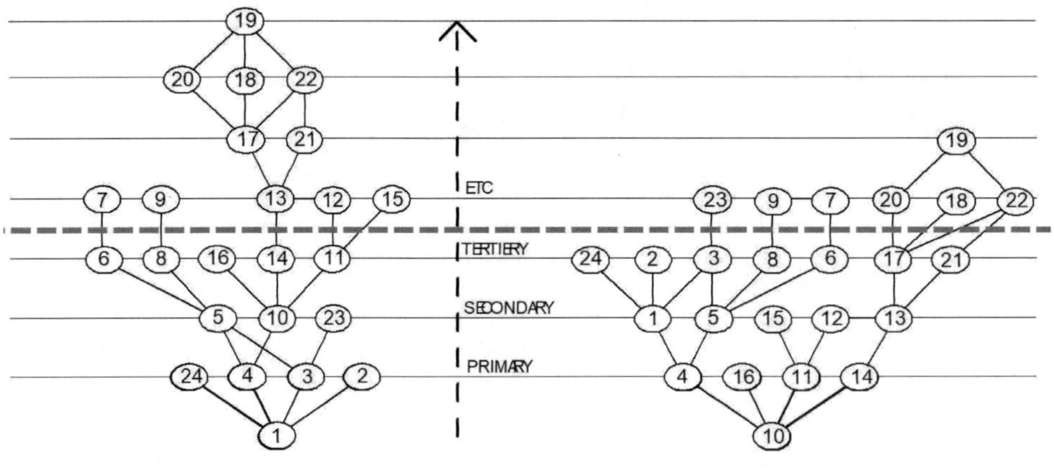

Note: Radius of analysis is set at a topological distance of 3. Source: Author

 https://doi.org/10.52202/067871-0004

Figure 13

Example of Closeness

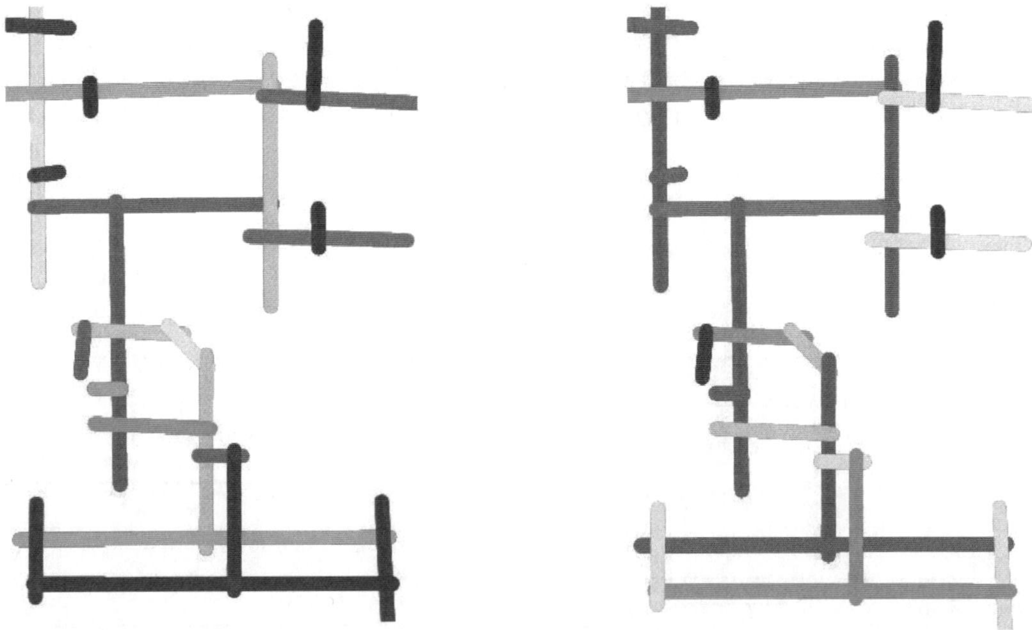

Note: Closeness-n (left) and Closeness-3 (right) values of the same layout. Source: Author

5. Variables

The final decision regarding Space Syntax analysis would be selection of variables that would be computed for the unit spaces (see Figure14). The variables 'Closeness' and Betweenness' have been explained before[3]. Researchers need to be very sure about their choice of both the variables and their justifications before any analysis and empirical data collection begin.

Figure 14

Example of Metric Betweenness and Closeness

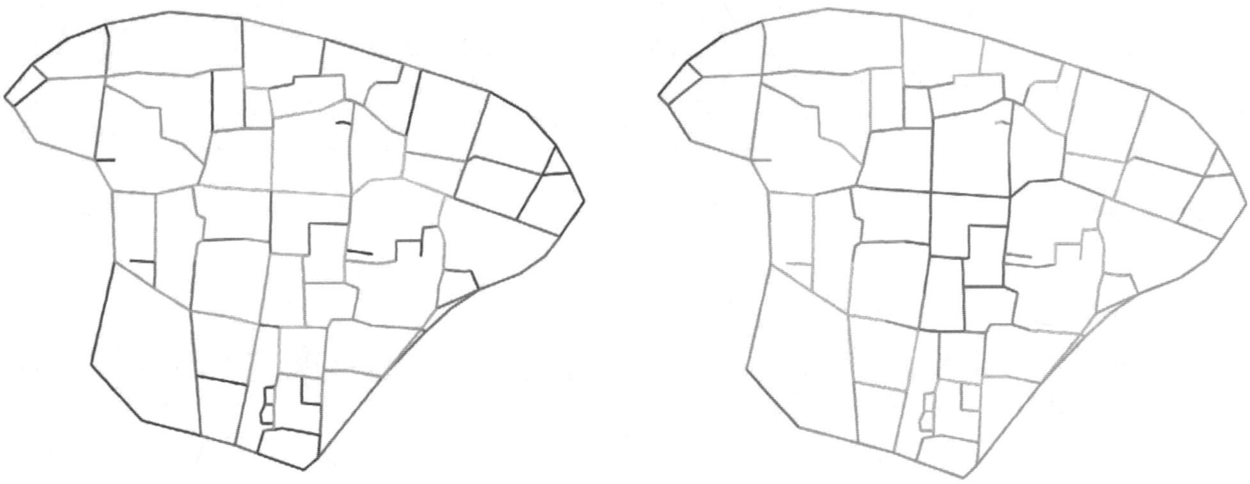

Note: Metric Betweenness-n (left) and Metric Closeness-n (right) modelled using segmented lines. Source: Author

[3]There are other variables, but these two are mostly used.

https://doi.org/10.52202/067871-0004

Conclusion

This paper began with some fundamental assumptions regarding Space Syntax, its relationships to human attributes, and its analytical techniques. These have developed and expanded over the years and Space Syntax analysis now can be undertaken in many different ways (see Table 1). In this regard, this paper has identified five very important categories about which the researcher needs to make thoughtful decisions, both in terms of the research questions being formulated, as well as the analyses that are conducted. It is vital to understand that the decisions regarding any one of these categories will produce changes in the computed values of each unit space and will have a profound effect on the results. These decisions are extremely important and must be clearly explained and related to the hypothesis of any research endeavor.

Table 1

Five Areas of Decision Making in Space Syntax Analysis

	DECISION AREAS	CHOICES AVAILABLE
	Unit Spaces	Axial Lines [All lines, Reduced Lines, Drawn Lines, Segmented Lines]
		Rooms
		Units of a grid
	Spatial System	Boundary
		What to include inside boundary
		Visibility and Accessibility require identification of different spatial systems.
	Concept of Distance between unit spaces, or connection types	Topological
		Geometric/ Angular
		Metric
	Radius of Analysis	Any topological steps (if topological)
		Any angle (if geometric)
		Any distance (if metric)
	Space Syntax Variables	Closeness (integration)
		Betweenness (Choice)

References

Alalouch, C., Aspinall, P., & Smith, H. (2009). On locational preferences for privacy in hospital wards. *Facilities, 27*(3/4, 2009), 88 - 106. https://doi.org/10.10.1108026327709109 33125

Alrashed, A. (2021). *Public experience of modern Saudi mega hospitals: A space syntax case study of King Fahad Specialist Hospital, Dammam*, Saudi Arabia. Texas Tech University. http://hdl.handle.net/2346/87466

Berhie, G. (2016). The effect of spatial configuration on land use and transport mode choices: *Space syntax exploration on gridded and non-gridded American cities*. [Dissertation Defense, Texas Tech University]. http://hdl.handlenet/2346/68040

Chambers, E. C., Bafna, S., & Machry, H. (2018). The association between apartment layout and depressive symptomology among Hispanic/Latino residents in low-income housing: The AHOME Study. *Journal or Urban Health, 95*(1), 51-60. https://doi.org/10.1007/s11524-017-0216-4

 https://doi.org/10.52202/067871-0004

Dalton, R. C. (2003). The decret is to follow your nose: Route path selection and angularity. *Environment and Behavior, 35*(1), 107-131. https://doi.ort/10.1177/00139165022238867
Haq, S. (1999). Can space syntax predict environmental cognition? *2nd International Space Syntax Symposium.*Brazilia, Brazil.
Haq, S., & Zimring, C. (2001). Just down the road a piece: The development of topological knowledge in building layouts. *Space Syntax: 3rd International symposium.* Ann Arbor, MI.
Hendrich, A., Chow, M. P., Bafna, S., Choudhary, R., Heo, Y., & Skierczynski, B. A. (2009). Unit-related factors that affect nursing time with patients: Spatial analysis of the time and motion study. *Health Environments Research and Design Journal, 2*(2), 5-20. https://doi.org/10.1177/193758670900200202
Hillier, B. (Ed.). (1970). *Architectural psychology.* RIBA publications.
Hillier, B. (1973). In defence of space. *RIBA Journal*, November, 539-544.
Hillier, B. (1977). A state of mind. *RIBA Jnl, Jubilee Edition*(May).
Hillier, B. (1999). The hidden geometry of deformed grids: Or, why space syntax works, when it looks as though it shouldn't. *Environment and Planning B, 26*(2), 169-191. https://doi.org/10.1068/b4125
Hillier, B. (2005). The art of place and the science of space. *World Architecture, 96-102.*
https://discovery.ucl.ac.uk/id/eprint/1678
Hillier, B. (2007). *Space is the machine: A configurational theory of architecture. Space Syntax.* UCL.
Hillier, B., Burdett, R., Peponis, J., & Penn, A. (1987). Creating life: Or, does architecture determine anything? *Architecture and Behavior/Architecture et Comportment, 3*(3), 233-250.
Hillier, B., & Hanson, J. (1984). The social logic of space (Paperback Edition 1988 ed.). *Cambridge University Press.*
Hillier, B., & Leaman, A. (1974). How is design possible? *Journal of Architectural Research, 3*(1), 4-11.
Hillier B., J, M., & P., O. S. (1972). Knowledge and design. *Environmental Design Research and Practice,* EDRA 3.
Hölscher, C., Brösamle, M., & Vrachliotis, G. (2012). Challenges in multilevel wayfinding: A case study with the space syntax technique. *Environment and Planning B: Planning and Design, 39*(1). https://doi.org/10.1068/b34050t
Kubat, A. S. (2009). The analysis of urban features that affect land values in residential areas. *7th International Space Syntax Symposium.* Stockholm, Sweden, KTH.
Lu, Y., & Zimring, C. (2012). Can intensive care staff see their patients? An improved visibility analysis methodology. *Environment and Behavior, 44*(6), 861-876. https://doi.org/10.1177/0013916511405314
Ossmann, M. (2016). *Exploring spatial risk: The impact of visibility on ICU mortality [PhD*, Georgia Institute of Technology].
Penn, A. (2001). Space syntax and spatial cognition Or, why the axial line. *Space Syntax 3rd International Symposium. GeorgiaTech*, Atlanta.
Penn, A., Hillier, B., Banister, D., & Xu, J. (1998). Configurational modelling of urban movement networks. *Environment and Planning B, 25*(1), 59-84. https://doi.org/10.1068/b250059
Peponis, J., Hadjinikolaou, E., Livieratos, C., & Fatouros, D. A. (1989). The spatial core of urban culture. *Ekistics, 56*, 334, 335 Jan-Feb, Mar-Apr, 43-55.
Peponis, J., Zimring, C., & Choi, Y. K. (1990). Finding the building in wayfinding. *Environment and Behavior, 22*(5), 555-590. https://doi.org/10.1177/0013916590225002
Saussure, F. D. (2006). *Writings in general Llinguistics* (C. Sanders & M. Pire, Trans.; S. Bouquet & R. Engler, Eds.). Oxford University Press.

Overview and Methodological Implications of Studying Virtual Nature for Human Health

Matthew H. E. M. Browning [a*] *and* ***Fu Li*** [a] *([a] Virtual Reality & Nature Lab, Department of Parks, Recreation & Tourism Management, Clemson University)*
**corresponding author*

ABSTRACT

Nature-based virtual reality (VR) experiences can promote mental health. We provide a primer on these experiences for environmental design researchers and practitioners. We begin with a review of reviews on existing nature-based VR experience research. We also include the prevailing research methods for studying these experiences. Next, we summarize the evidence for these experiences impacting different mental health outcomes. We proceed with an overview of how to create and playback these experiences in VR systems. Several research needs are then proposed, including studying populations of greatest need for these experiences and studying a variety of natural landscapes. We conclude with six lessons and tips based on our 12 years of VR research. These include enhanced recruitment methods, incentive distribution, participant onboarding, individual difference consideration, scientific best practice alignment, and scope of research breadth. Pursuing these tips will help improve our understanding of the efficacy and administration of nature-based VR experiences for promoting mental health.

INTRODUCTION

The demand for at-home healthcare systems rose sharply during the COVID-19 pandemic. Mental health also declined from fear, infections and mortality, financial concerns, caregiver demands, and social constraints. This union warranted more research on relatively low-cost, mobile, self-care mental health interventions. Modern virtual reality (VR) technology with restorative nature imagery presents promising delivery mechanisms for these interventions (Riva et al., 2020).

VR has existed for decades, but recent advances have allowed head-mounted displays (HMDs) to be accessible and affordable (Anthes et al., 2016; Mehrfard et al., 2020). Users can undertake a variety of clinical interventions, including distraction therapy, mindfulness training, cognitive behavior therapy, guided imagery, and attention/stress reduction from exposure to restorative natural environments (White et al., 2018). Virtual environments can be created with modest cost and skill using 360-degree cameras or more expensive and immersive approaches. We present a state-of-knowledge overview of VR devices, the efficacy of VR interventions for different outcomes, approaches to VR content creation, research needs, and methodological recommendations to enable environmental research and design scholars to use this technology.

 https://doi.org/10.52202/067871-0005

State-of-Knowledge

Understanding past research on restorative natural environments in VR is made possible by several systematic reviews. Browning et al. (2020) identified nearly 148 articles published by December 2018 that used some form of simulated nature. Among these, 25 experiments used VR head-mounted displays (HMDs), and one experiment used a room-scale environment known as a cave automatic virtual environment (CAVE) (Figure 1). Frost et al. (2022) found 21 studies of nature imagery in VR and its effects on psychological well-being that were published between 2015 and July 2020. These studies included 19 experiments with HMDs and two with CAVEs. Nukarinen et al. (2021) found 21 articles on the restorative effects of nature-based imagery in VR. Most recently, a preprint by Spano et al. (2022) listed 38 articles on virtual nature that were published between 1996 and September 2021. Collectively, these reviews suggest that approximately one-third of past research has tested the effects of VR among student populations, and another third has involved clinical populations, such as patients with dementia or mood disorders. Sample sizes have ranged widely from 14 to 1280, and exposures have commonly lasted 5 to 10 min. A striking finding is the lack of longitudinal or repeated measures designs in past research, whereby multiple exposures are requested of participants, and longer-term effects are recorded from exposure. Only a few studies have used such designs with 10 min doses across 10 days (Veling et al., 2021), 20 min viewings across 7 days (Lakhani et al., 2020), or 13 min viewings on average across 7 days (Chin et al., 2022; Reynolds et al., 2022).

Figure 1

HMD (a) and CAVE (b) VR Systems

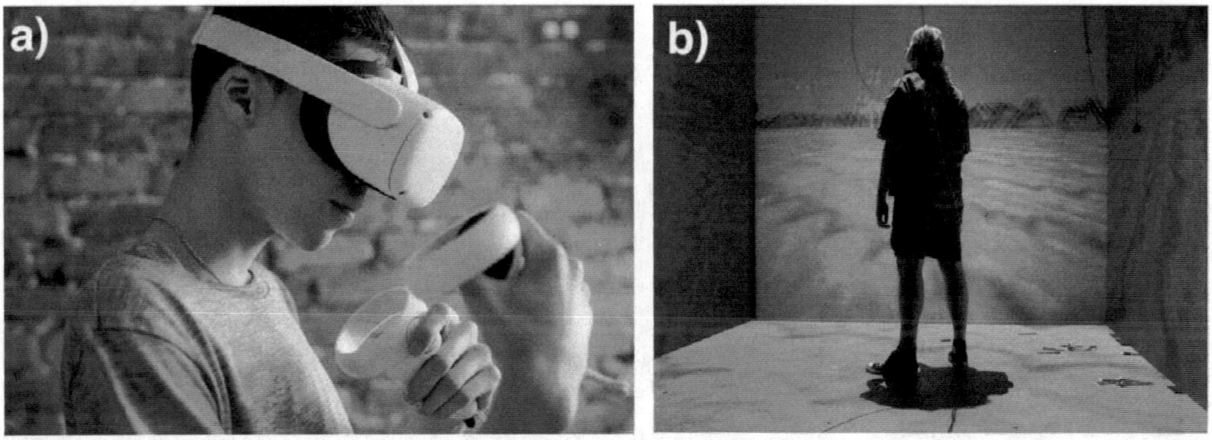

Note: Source (Figure-a) from Pexels (https://www.pexels.com/photo/man-in-vr-headset-testing-technology-7241503/) and Source (Figure-b) Wikimedia (https://commons.wikimedia.org/wiki/File:CAVE_Crayoland.jpg). Public Domain.

Health Impacts

Nature imagery in VR generally has positive effects on users, but the consistency of effects varies by outcome. For instance, the effects on positive emotions are inconsistent (Browning, Shipley, et al., 2020; Spano et al., 2022). Effects on negative emotions are more robust; most studies have found drops in stress, anger, tension, depressive or anxiety symptoms, fatigue, and physiological measures (i.e., skin conductance). Only a small number of studies have found non-significant changes in these outcomes. At least four studies have examined alpha brain waves to measure relaxation but reported mixed effects. Perceived restorativeness has generally been higher for virtual natural environments than for virtual built environments (Browning, Saeidi-Rizi, et al., 2020). The therapeutic restorative effects of virtual nature are

https://doi.org/10.52202/067871-0005

likely to act through multiple dimensions, including physiological, affective, cognitive, social, ecological, and transcendent restoration (Nukarinen et al., 2021). There is insufficient research and evidence for the beneficial effects of exposure on cognitive resources or attention.

Content Creation

The sources of content creation for virtual nature are multifaceted. Eventually, all types of VR content are presented in 360-degree spherical or three-dimensional (3D) formats. Consumer and professional panoramic cameras (i.e., Insta 360, Ricoh Theta, or Nokia OZO) can capture and stitch together 360-degree pictures/video (Browning, Suppakittpaisarn, et al., 2020; Joseph et al., 2020). Drones and handheld devices can scan and reconstruct 3D environments using apps like 3D Scanner for Apple iPhone, PIX4D or Metashape for drones, or most software that accompanies cameras with differential GPS or LIDAR (F. Li & Chang, 2019). Alternatively, environments can be built with purchased, self-built or free assets in gaming engines (i.e., Autodesk Revit/3DMax, SketchUp, Blender, Unity3D, and Unreal) (Joseph et al., 2020; Yin et al., 2019). Any of these content types can be uploaded to HMD or CAVE VR systems for playback to provide immersive experiences. With the maturity of 3D creator tools and control devices, scholars can increasingly produce high-quality computer-generated imagery (CGI) environments that provide stationary, interactive or mobility-based experiences, such as walking, biking, or driving (Figure 2) (Calogiuri et al., 2022; Jiang et al., 2020, 2021; Li et al., 2022; Rockstroh et al., 2020).

Figure 2
Examples of Production Streams for Stationary, Interactive, and Mobility-based Virtual Nature Experiences

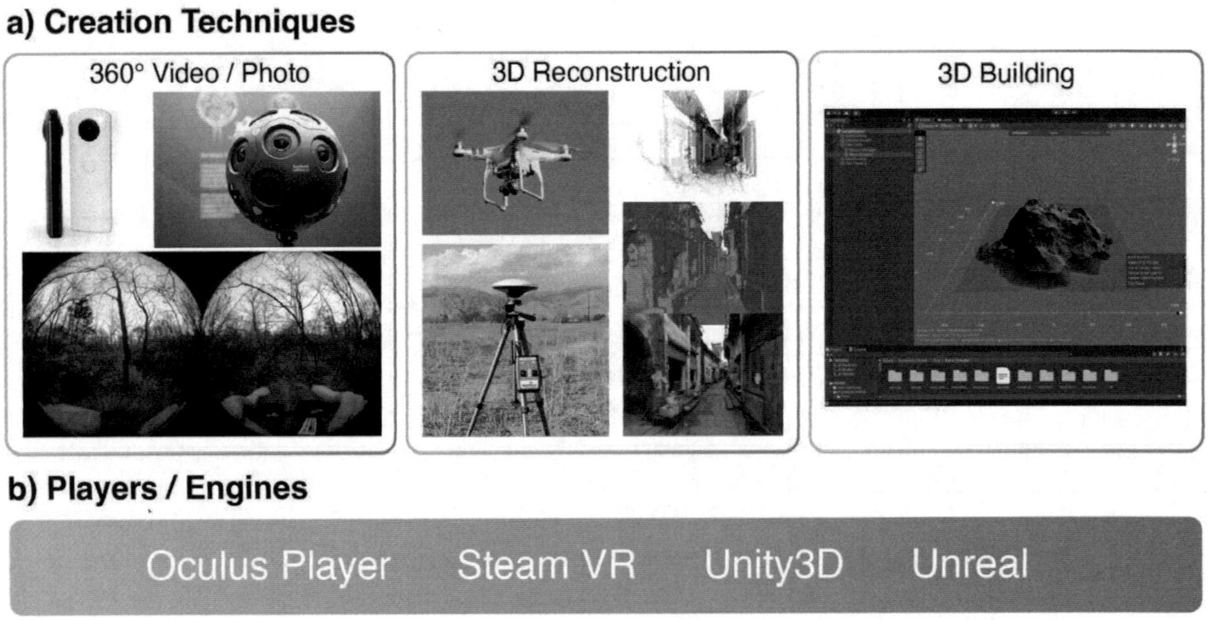

 https://doi.org/10.52202/067871-0005

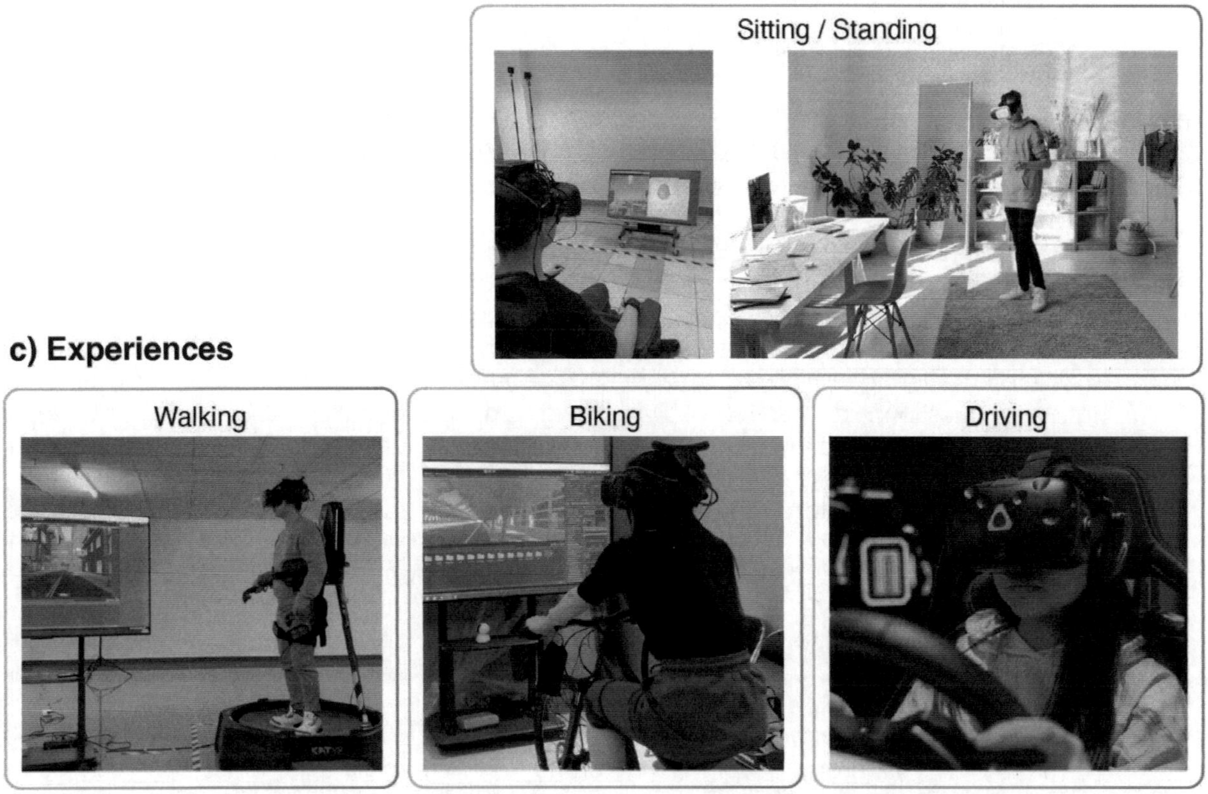

Note: a) Virtual reality content can be created by 360-degree camera capture, 3D scanning and reconstruction, or 3D building; b) Appropriate engines must be selected to play/stream the content on the headset; c) Different input devices can move users in virtual environments to provide a more realistic experience.
Illustrated by the author. Contents from Wikimedia, Pixbay, Higher Education Press (China), Pexels, and author. Proper authorization for use has been obtained.

Research Needs

Studies on virtual nature should continue to expand their samples to populations of greatest need for these interventions. Populations receiving in-patient care could benefit from virtual exposure to nature since they are unlikely to have window views of nature, indoor plants, or outdoor nature-rich environments (White et al., 2018). Recruitment does not need to be limited to the conditions studied previously; patients on maternal bedrest (Browning & Lee, 2011) or recovering from surgery (Ulrich, 1984) and healthcare personnel working long hours (Jiang et al., 2022) are also relevant. Military personnel, the incarcerated, and civilian employees working in isolated, confined, and extreme (ICE) environments may further benefit from virtual nature exposure. Such possibilities have prompted a line of research on VR interventions in space, on submarines and cargo ships, and in polar locations, though rigorous experimental research is lacking (Abbott & Diaz-Artiles, 2022; Anderson et al., 2017, 2022; Lockard & Kaufman, 2019; Neilson et al., 2021).

Defining what types of nature impact health outcomes among different populations should also be investigated. People familiar with deserts may benefit as much from virtual exposure to that landscape as an urban park with trees (Yin et al., 2022). Natural landscapes consist of water, plants, and/or rocks and minerals (Li et al., 2022). Most research on the health benefits of nature exposure has focused on plant-rich environments ("greenspace"), with an increasing number of studies on liquid water landscapes such as oceans, lakes, rivers, and canals ("bluespace"). Natural landscapes are not always green or blue, and the effects of other landscapes are worth attention, such as deserts, caves, and areas with abundant solid-state water (ice or snow).

Methodological Implications

Based on our team's 12 years of experience conducting VR research, we highlight some lessons and tips we have learned regarding the methodological design and conduct of studies on virtual nature:

- **Develop better recruitment methods** - some middle/older adults are disinterested in VR or concerned about becoming sick, which causes a barrier to recruitment. We have not yet observed effective strategies to encourage use among this population.
- **Provide regular and predictable incentives** - like other longitudinal research, VR experiments have the challenge of high experimental mortality (i.e., participant dropout). Compliance with multiple questionnaires can be as challenging as maintaining adherence to a VR intervention. Providing incentives as a study progresses rather than a lump sum at the end of a study may encourage questionnaire compliance.
- **Onboard participants through training videos and calls** – VR headsets can be mailed to participants with prepaid shipping for returns. YouTube or similar cloud-based videos that show how to put on the headset, what actions the buttons perform, and what participants will see in the headset appear to be a convenient method to train participants to use VR at home.
- **Consider individual differences** – people may respond differently to virtual nature. As such, demographic factors and potential psychological confounders should be measured, including childhood and adult experiences in nature and past VR use.
- **Follow scientific best practices** – clinical trials and experimental designs should be pre-registered (i.e., clinicaltrial.org, COS.io). Some journals are recommending or requiring manuscripts to include registration numbers with submissions. Furthermore, given the premise that nature experiences act through the restoration of a depleted state, experiments should focus more on pre-fatiguing participants' attentional capacities or inducing acute stress. (Browning, Shipley, et al., 2020; Hartig, 2011, 2020).
- **Expanding the topic** – studying human-environment interactions is complex. We call for investigations of a broad range of topics surrounding virtual nature, such as the restorative potential of multi-sensory experiences, differential effects across cultural contexts, and collateral impacts on other societally relevant outcomes such as climate change concern and pro-environmental behaviors. We also acknowledge disparities in access to health-promoting environments outside of VR. Ignoring these disparities by recommending virtual nature might extend historic environmental injustices. Research on how VR can allow policymakers and designers to witness the negative health impacts of nature-poor environments so they prioritize overcoming these real-world problems is an important area of future research.

Conclusion

VR provides access to restorative natural environments in situations where physical access is limited. Research on this topic has grown substantially in the last 20 years. Our current understanding of the beneficial health effects of virtual nature is limited by the relatively narrow populations studied and lack of longitudinal research. We encourage future research on this topic with rigorous study designs that respond to these limitations and other methodological considerations regarding recruitment, incentives, onboarding, individual differences, and participants' baseline states for the outcomes of interest.

 https://doi.org/10.52202/067871-0005

Acknowledgments

This paper summarizes the expertise and knowledge based on years of research in two laboratories: the Virtual Reality & Nature Lab (VRN) and the Digital Technology and Geo-Design Lab (DTGDL). The VRN is located at Clemson University and was formerly based at the University of Illinois at Urbana-Champaign, USA. The lab is led by Prof. Matthew H. E. M. Browning and managed by Olivia McAnirlin. The DTGDL is a Lab at the College of Architecture and Landscape Architecture, Peking University, China, and is led by Dr. Liyan Xu and Fu Li.

The authors would like to thank the many lab members, affiliates, and research participants. Without their efforts, the expertise summarized in this paper would not have been possible.

References

Abbott, R. (Woodruff), & Diaz-Artiles, A. (2022). The impact of digital scents on behavioral health in a restorative virtual reality environment. *Acta Astronautica, 197*, 145–153. https://doi.org/10.1016/j.actaastro.2022.05.025

Anderson, A. P., Mayer, M. D., Fellows, A. M., Cowan, D. R., Hegel, M. T., & Buckey, J. C. (2017). Relaxation with immersive natural scenes presented using virtual reality. *Aerospace Medicine and Human Performance, 88*(6), 520–526. https://doi.org/10.3357/amhp.4747.2017

Anderson, A. P., Stankovic, A., Cowan, D., Fellows, A., & Buckey, J. (2022). Natural scene virtual reality as a behavioral health countermeasure in isolated, confined, and extreme environments: Three isolated, confined, extreme analog case studies. *Human Factors: The Journal of the Human Factors and Ergonomics Society*, 001872082211006. https://doi.org/10.1177/00187208221100693

Anthes, C., Garcia-Hernandez, R. J., Wiedemann, M., & Kranzlmuller, D. (2016). *State of the art of virtual reality technology, 1–19*. https://doi.org/10.1109/aero.2016.7500674

Browning, L. M., & Lee, S. (2011). Pregnancy and place: Creating therapeutic gardens for maternity care patients. *Journal of Therapeutic Horticulture,21*(2) 42-67.

Browning, M. H. E. M., Saeidi-Rizi, F., McAnirlin, O., Yoon, H., & Pei, Y. (2020). The role of methodological choices in the effects of experimental exposure to simulated natural landscapes on human health and cognitive performance: A systematic review. *Environment and Behavior, 7*(1), 1–43. https://doi.org/10.1177/0013916520906481

Browning, M. H. E. M., Shipley, N., McAnirlin, O., Becker, D., Yu, C.-P., Hartig, T., & Dzhambov, A. M. (2020). An actual natural setting improves mood better than its virtual counterpart: A meta-analysis of experimental data. *Frontiers in Psychology, 11*, 2200. https://doi.org/10.3389/fpsyg.2020.02200

Browning, M., Suppakittpaisarn, P., Jiang, S., & Joseph, A. (2020). Human health assessments of green infrastructure designs using virtual reality. *Landscape Architecture, 27*(9), 35–49. https://doi.org/10.14085/j.fjyl.2020.09.0035.15

Calogiuri, G., Keegan, B. J., Birkheim, S. L., Rydgren, T. L., Flaten, O. E., Fröhlich, F., & Litleskare, S. (2022). A mixed-methods exploration of virtual reality as a tool to promote green exercise. *Scientific Reports, 12*(1), 5715. https://doi.org/10.1038/s41598-022-09622-x

Chin, S., Cavadino, A., Akroyd, A., Tennant, G., Dobson, R., Gautier, A., & Reyonds, L. (2022). An investigation of virtual reality nature experiences in patients With metastatic breast cancer: Secondary analysis of a randomized controlled trial. *JMIR Cancer, 3*(8). https://doi.org/10.2196/38300

Frost, S., Kannis-Dymand, L., Schaffer, V., Millea, P., Allen, A., Stallman, H., Mason, J., Wood, A., & Atkinson-Nolte, J. (2022). Virtual immersion in nature and psychological well-being: A systematic literature review. *Journal of Environmental Psychology, 80*, 1-9. https://doi.org/10.1016/j.jenvp.2022.101765

Hartig, T. (2011). Issues in restorative environments research: Matters of measurement. *Psicologia Ambiental, 41-66*.

Hartig, T. (2021). Restoration in nature: Beyond the conventional narrative. In A. R. Schutte, J. Torquati, & J. R. Stevens (Eds.), *Nature and psychology: Biological, cognitive, developmental, and social pathways to well-being* (pp. 89–151). Springer Nature.

Jiang, B., He, J., Chen, J., & Larsen, L. (2021). Moderate is optimal: A simulated driving experiment reveals freeway landscape matters for driving performance. *Urban Forestry & Urban Greening, 58*, 126976. https://doi.org/10.1016/j.ufug.2021.126976

Jiang, B., He, J., Chen, J., Larsen, L., & Wang, H. (2020). Perceived green at speed: A simulated driving experiment raises new questions for attention restoration theory and stress reduction theory. *Environment and Behavior, 53*(3), 296-335. https://doi.org/10.1177/0013916520947111

Jiang, S., Allison, D., & Duchowski, A. T. (2022). Hospital greenspaces and the impacts on wayfinding and spatial experience: An explorative experiment through immersive virtual environment (IVE) techniques. *HERD: Health Environments Research & Design Journal, 15*(3), 206-228. https://doi.org/10.1177/19375867211067539

Joseph, A., Browning, M. H. E. M., & Jiang, S. (2020). Using immersive virtual environments (IVEs) to conduct environmental design research: A primer and decision framework. *HERD: Health Environments Research & Design Journal, 13*(3), 11–25. https://doi.org/10.1177/1937586720924787

Lakhani, A., Martin, K., Gray, L., Mallison, J., Grimbeek, P., Hollins, I., & Mackareth, C. (2020). What is the impact of engaging with natural environments delivered via virtual reality on the psycho-emotional health of people with spinal cord injury receiving rehabilitation in hospital? Findings from a pilot randomized controlled trial. *Archives of Physical Medicine and Rehabilitation, 101*(9), 1532–1540. https://doi.org/10.1016/j.apmr.2020.05.013

Li, F., & Chang, H. (2019). UAV measurement in landscape architecture research and practice. *Landscape Architecture Frontiers, 7*(2), 38. https://doi.org/10.15302/j-laf-20190204

Li, H., Browning, M. H. E. M., Rigolon, A., Larson, L. R., Taff, D., Labib, S. M., Benfield, J., Yuan, S., McAnirlin, O., Hatami, N. & Kahn, P. H. (2022). Beyond "bluespace" and "greenspace": A narrative review of possible health benefits from exposure to other natural landscapes. *Science of The Total Environment*, 159292. https://doi.org/10.1016/j.scitotenv.2022.159292

Li, H., Zhang, X., Wang, H., Yang, Z., Liu, H., Cao, Y., & Zhang, G. (2021). Access to nature via virtual reality: A mini-review. *Frontiers in Psychology, 12*, 725288. https://doi.org/10.3389/fpsyg.2021.725288

Lockard, E., & Kaufman, A. (2019, July 7-11). *Bringing nature into space: The restorative potential of virtual environments for long-term space travel.* [Conference paper presentation]. 49th International Conference on Environmental Systems, Boston, MA, United States.

Mehrfard, A., Fotouhi, J., Taylor, G., Forster, T., Navab, N., & Fuerst, B. (2020). *A comparative analysis of virtual reality head-mounted display systems*. Arxiv.Org. arxiv.org

Neilson, B. N., Craig, C. M., Altman, G. C., Travis, A. T., Vance, J. A. & Klein, M. I. (2021). Can the biophilia hypothesis be applied to long-duration human space flight? A mini-review. *Frontiers in Psychology, 12*, 703766. https://doi.org/10.3389/fpsyg.2021.703766

Nukarinen, T., Rantala, J., Korpela, K., Browning, M. H. E. M., Istance, H. O., Surakka, V., & Raisamo, R. (2021). Measures and modalities in restorative virtual natural environments: An integrative narrative review. *Computers in Human Behavior*, 107008. https://doi.org/10.1016/j.chb.2021.107008

Reynolds, L. M., Cavadino, A., Chin, S., Little, Z., Akroyd, A., Tennant, G., Dobson, R., Broom, R., & Gautier, A. (2022). The benefits and acceptability of virtual reality interventions for women with metastatic breast cancer in their homes: A pilot randomised trial. *BMC Cancer, 22*(1), 360. https://doi.org/10.1186/s12885-021-09081-z

Riva, G., Bernardelli, L., Browning, M., Castelnuovo, G., Cavedoni, S., Chirico, A., Cipresso, P., Paula, D. M. B. de, Lernia, D. D., Fernández-Álvarez, J., Figuerás, N., Fuji, K., Gaggioli, A., Gutierrez-Maldonado, J., Hong, U., Mancuso, V., Mazzeo, M., Molinari, E., Moretti, L., … Wiederhold, B. K. (2020). COVID feel good - An easy self-help virtual reality protocol to overcome the psychological burden of Coronavirus. *Frontiers in Psychiatry*. https://doi.org/10.3389/fpsyt.2020.563319

Rockstroh, C., Blum, J., Hardt, V., & Göritz, A. S. (2020). Design and evaluation of a virtual restorative walk with room-scale virtual reality and impossible spaces. *Frontiers in Virtual Reality, 1*, 598282. https://doi.org/10.3389/frvir.2020.598282

 https://doi.org/10.52202/067871-0005

Spano, G., Theodorou, A., Reese, G., Carrus, G., Sanesi, G., & Panno, A. (2022). Virtual nature and psychological outcomes: A systematic review. *OSF Preprints*. https://doi.org/https://doi.org/10.31219/osf.io/8ux9a

Ulrich, R. S. (1984). View through a window may influence recovery. *Science, 224*(4647), 420–421. https://doi.org/10.1126/science6143402

Veling, W., Lestestuiver, B., Jongma, M., Hoenders, H. J. R., & Driel, C. van (2021). Virtual reality relaxation for patients with a psychiatric disorder: Crossover randomized controlled trial. *Journal of Medical Internet Research, 23*(1), e17233. https://doi.org/10.2196/17233

White, M. P., Yeo, N. L., Vassiljev, P., Lundstedt, R., Wallergård, M., Albin, M., & Lõhmus, M. (2018). A prescription for "nature" - the potential of using virtual nature in therapeutics. *Neuropsychiatric Disease and Treatment, 14*, 3001–3013. https://doi.org/10.2147/ndt.s179038

Yin, J., Arfaei, N., MacNaughton, P., Catalano, P. J., Allen, J. G., & Spengler, J. D. (2019). Effects of biophilic interventions in office on stress reaction and cognitive function: A randomized crossover study in virtual reality. *Indoor Air, 29*(6), 1028–1039. https://doi.org/10.1111/ina.12593

Yin, J., Bratman, G. N., Browning, M. H. E. M., Spengler, J. D., & Olvera-Alvarez, H. A. (2022). Stress recovery from virtual exposure to a brown (desert) environment versus a green environment. *Journal of Environmental Psychology, 81*, 101775. https://doi.org/10.1016/j.jenvp.2022.101775

https://doi.org/10.52202/067871-0005

Hindsight 20/20: Learning from a 2020 National Study of the Outpatient Experience

Nicholas Watkins, *PhD (Gensler)*, ***April Spivack***, *PhD (Hanken School of Economics)*, ***Kyle Sellers***, *MA (Gensler), and* ***Anita Grabowska***, *MEd (Gensler)*

ABSTRACT

Pandemic-related, economic, policy, and technological forces have only amplified volatility in a United States healthcare industry where margins are shrinking and patient consumerism spells success or failure. However, investments in innovative outpatient strategies have been paying off by affording greater value to consumers. The outpatient care bellwether, with its flurry of service and facility hybrids, and the innovations emerging from the COVID-19 pandemic create an opportunity for us to take a comprehensive snapshot of the state-of-the industry and understand how social determinants of health, design, and digital design shape outpatient experiences and outcomes. This nationwide survey of 3,578 outpatients represents over 83 diagnoses, 81 health systems, and 16 outpatient settings. The study tested the themes of outpatient engagement, perceived care quality, consumer loyalty, pandemic experience, and virtual visit quality; identified outpatient populations by their demographic characteristics and their experiences with and access to outpatient care; identified design and digital design features that positively impact outpatient experience and outcomes; and explored optimal mixes of outpatient services. Opportunities and unexpected challenges of the pandemic, questionnaire development, sampling, data collection, and analyses will be discussed. Audience members will contextualize the methods and findings to their own experiences and understand ways to position themselves for success when performing similar research.

CONTEXT

The Gensler Research Institute (GRI) has a history of performing nation-wide and international panel-based self-report surveys, starting with its UK Workplace Survey conducted in 2005. Since then, the GRI's mandate has grown to include a firmwide research grants program, mixed methods research conducted across several industries, and media development and oversight including the Gensler Research Catalog and the Gensler Research Institute website (Gensler Research Institute, 2022).

More recent responsibilities include the development and maintenance of the GRI Experience Index (EXI) series of self-report questionnaires. The EXIs measure consumer experiences including those for life sciences, mixed-use properties and neighborhoods, sports, urban planning, education, retail settings, hospitality, climate change and resilience, and others. Begun in 2019, the US Outpatient Experience Index (O-EXI) was the inaugural healthcare-related addition to the EXI series.

The O-EXI was the first-of-its-kind, a nationwide survey in focus and depth. Similar to the Press Ganey Satisfaction Survey or the Center of Medicare & Medicaid Services (CMS) Consumer Assessment of Healthcare Providers & Systems (CAHPS) Series of surveys, the O-EXI focuses on the patient experience. However, unlike CAHPS or Press Ganey, the O-EXI has a deep focus on the patient experience with outpatient healthcare design and operations. What follows are key takeaways and lessons learned from the development and implementation of the O-EXI. Many points here can be extended to surveys for other industries beyond healthcare.

 https://doi.org/10.52202/067871-0006

Online Survey Panel Providers: Criteria for Selection

The choice of panel provider is the most critical step in planning an online panel-based survey and may be an experience unfamiliar to most design researchers. Therefore, this selection is worth detailed attention in this paper.

Overall, the choice of online panel provider should be informed by several parameters including the project's budget and duration, the size of the client researcher's team and its skillsets, the quality of the provider's panel recruitment and management, and the provider's partnerships and professional affiliations. The choice can be challenging; strict standards for the transparency of panel providers' panel recruitment and management techniques are difficult to access and hard to regulate (Sandorf et al., 2022).

Online Panel Provider Types

Online panel services include offerings from three sources: crowdsourced and largely self-serviced platforms such as Amazon Mechanical Turk and Prolific; professional panel providers such as Kantar Profiles or Ipsos who maintain proprietary panels of panelists; and panel aggregators such as Qualtrics or OpinionRoute who pull from several partnering panels. The professional panel providers and the panel aggregators can offer easily accessible bundles of offerings including questionnaire development, questionnaire programming, survey administration, data processing and data visualization, among others. The majority of panel providers offer incentives to their panelists who complete their surveys. Typically, these incentives are in the form of points that the panelists can convert to online gift cards, cash deposits, and charity donations.

Amazon Mechanical Turk (AMT) is a crowdsourced and self-service panel platform. It is increasingly popular among advertising and social science researchers. It is cost effective and can produce a quality of data similar to that of professional online panel providers if quality control options are taken (Endres et al., 2022; Sheehan & Pittman, 2016). For example, with AMT, a client researcher can engage prescreened/vetted panelists and specify panelists with a high reputation rating. A researcher can withhold incentives for and give a poor approval rating to AMT panelists who do not meet the criteria for a successfully completed survey (crowdsourcing platforms and ways to improve data quality are discussed elsewhere in these proceedings by A. S. Devlin).

However, AMT was not considered an optimal fit when the O-EXI was under development. Instead, Kantar Profiles, a professional online panel provider was engaged for several reasons. First, Kantar Profiles' specialty panels of patients and care providers are profiled to represent several diverse and important demographic characteristics that influence experiences with healthcare including lifestyle habits (i.e., smoking, exercise), diagnosis, and comorbidities. AMT tends to represent the general population (Robinson et al., 2019). Second, credible estimates on the size of AMT's pool of panelists vary but it may in fact not exceed more than 250,810 individuals at any one time (Robinson et al., 2019). Currently, Kantar Profiles' LifePoints panel has over 900,000 panelists in the US (B. Cash, Personal Communication, September 22, 2022).

Third, of the AMT sample a small percentage of "superworkers" are continually recycled. These superworkers complete many more surveys on AMT than other panelists who are usually considered ineligible, lack the technology to secure desirable surveys, or do not have a strong commitment to earning money from the AMT platform (Robinson et al., 2019). On the other hand, Kantar Profiles continuously runs recruitment campaigns across several media channels and sources to ensure it acquires diverse and unbiased panelists. The panelists are pre-screened for past survey and panel participation. Panelists' profiles are continuously and automatically updated based on survey responses. These regular updates reflect any changes to residence or lifestyle (i.e., exercise, smoking cessation, diet, etc.). Finally, regular audits ensure that poor-performing panelists are detected and then removed (Kantar Profiles, 2022).

Fourth, Kantar Profiles had a strong reputation. Kantar Profiles was recommended by prominent market researchers in healthcare systems and its employees regularly contribute to scientific journals. Finally, with several nation-wide and international surveys in progress within the GRI, it was important to engage with a professional and trusted online panel provider who could share project management responsibilities.

https://doi.org/10.52202/067871-0006

Industry Accepted Criteria and Certification

Fortunately, there are criteria and certifications by which to evaluate an online panel provider. The European Society for Opinion and Market Research (ESOMAR) recommends asking 36 specific questions of an online panel provider to ensure it meets the quality and standards of a client researcher (ESOMAR, 2021). In fact, several providers can direct you to their answers to the ESOMAR questions on their websites. In addition to ESOMAR, it is useful to find out if the online panel provider is International Standards Organization (ISO) 20252 certified. ISO 20252 covers quality standards for planning, conducting, supervising, and reporting out to clients by online panel providers, including standards for project management and data security (ISO, 2019).

Representativeness of Sample

If the goal is to generalize a survey's findings to a region or country, then a client researcher will want to engage an online panel-provider whose panelists represent the population of interest. This representative sample will be obtained by the provider through an extensive recruitment process that includes random selection of panelists for survey participation. Random selection ensures everyone in the population has a known and nonzero probability of being recruited to the panel. Additionally, the sample is usually stratified by Census data to proportionally reflect different aspects of the population such as ethnicity or region. To further ensure representativeness of the sample, the panel provider should give free access to the Internet and hardware for panelists who normally would not take part in an online panel because they do not have the technology to answer an online survey.

Specialized Panels

Some online panel providers specialize in specific groups of panelists such as key opinion leaders (KOLs) in an industry, attorneys, homeowners, small businesses, specialty physicians, engineers, laboratory research scientists, construction workers, adults over the age of 65, and so on. Similarly, they can specialize in methodologies such as computer-assisted telephone interviews (CATIs) or personal interviewing (CAPI). The panel provider can recommend a method based on their experiences with the population of interest.

Prescreening of Panelists

The online panel provider should be able to share the prescreening/vetting methods they use to ensure quality and verification of panelists. For example, Kantar Profiles uses a variety of real-time checkpoints during panelist registration to its proprietary LifePoints panel and when a panelist starts an online survey. The address and zip code provided by a panelist can be compared to a local directory. Proxy detection can find a proxy server that is hiding a registrant's true IP address. E-mail verification can ensure the panelist's e-mail address is unique. Human-robot tests with CAPTCHAs are easy and highly effective at distinguishing between an actual panelist and a bot. Digital finger printing can detect duplicate entries in a survey. Kantar Health also uses a proprietary double opt-in process where individuals must first provide their demographic information in a registration survey (single opt-in point), pass through data quality checkpoints, agree to their country specific privacy policies, and finally confirm their email address (the double opt-in point).

All of these and other quality controls will be especially important when choosing a panel aggregator. To ensure the integrity of the survey's sample, the panel aggregator should hold their partner panels and any of their proprietary panels to similar and consistent panel quality controls. For example, it is useful to know if the panel aggregator uses social media to supplement panelist recruitment and meet their quotas instead of maintaining their universal prescreening standards.

Other Considerations

Currently, many online panel providers incentivize their panelists per capita, so the client researcher's budget needs to factor in the rate or cost the panel provider will charge for each completed survey by a panelist. Especially for targeted samples of panelists with rare and very specific characteristics, the provider will charge more. Rates seen by a co-author have been as high as $160 per a completed

 https://doi.org/10.52202/067871-0006

survey. Regardless of rate, a client researcher should be upfront with a panel provider that they should only charge for completed surveys and not for incomplete surveys or disqualified panelists. If at the end of a study a panel provider could not meet their obligation ("could not meet quota"), it should provide the client a credit or refund. This should also be reflected in the terms of the contract.

Besides any of the ESOMAR questions and ISO criteria, client researchers should come prepared to an initial discussion with a panel provider with the project budget, estimated sample size, sampling strategy, eligibility requirements for the survey, quotas that represent subgroups of panelists, estimated completion time of the survey by a panelist, and duration for data collection. Ideally, a client researcher will have a close-to-final if not final version of the survey with questions that disqualify ("screen out") panelists who would not qualify for the study; such questions are typically placed at the very beginning of the questionnaire. Often overlooked but crucial is to ask if the provider can support a longitudinal study if the client researcher has an opportunity to perform multiple waves of the survey. A panel provider will use all the previously mentioned aspects to estimate feasibility and cost of the survey and possibly factor in any discounts based on their impression of the client's skills, experience, and reputation. Cheaper is not better and the old adage, "you get what you pay for" does indeed apply, even if it is not explicitly stated.

Take the time and effort to solicit proposals from multiple online panel providers. Professional online panel providers are highly competitive and constantly evolve either with new offerings, techniques, or panelist types. For example, do not be surprised if a provider reports their panel and panel partners cannot with confidence support a survey of geneticists or that such a survey would far exceed your budget. However, in 2 or 3 years the same provider may report not only that it can field a survey of geneticists but that the entire survey can take less than 2 weeks at half the cost of the prior quote. If large professional panel surveys are routine for a client researcher, then regular attendance and participation in conferences hosted by organizations such as the American Association for Public Opinion Research (AAPOR) are ideal for the latest industry news (i.e., acquisitions, closings, new talent, transition of talent between companies, rates), networking, and brushing up on the latest methods, their advantages, and their limitations.

The chosen online panel provider will need to review the questionnaire the client researcher has developed and possibly even program it into a web-survey platform they routinely use. This review will ensure the survey is accessible and in a format that is familiar to the provider's panelists. However, a client researcher may have already programmed their survey into a web-survey platform such as Qualtrics or SurveyMonkey. If this is the case, check that the provider can accommodate the same platform(s). If not, the client researcher will need to manage and monitor quotas and response rates during data collection as the provider may not have the capability to do so. This additional responsibility can be easily overlooked during provider selection and project planning. It should be factored in the client researcher's work plan, budget, and staffing for the project and discussions with the provider.

Questionnaire Development, Programming, and Administration

It should be universally understood by an experienced design researcher in practice that extensive secondary research is essential to the development of any study and its questionnaire(s). Instead of a focus on secondary research, the co-authors recommend the reader consult the excellent research texts on the market with sections dedicated to secondary research (Devlin, 2020; Shepley, 2011; Sommer & Sommer, 2002). Instead, what follows are firsthand tips and applications that were essential to the quality of the O-EXI survey's development, programming, and administration. These may be less accessible or known to design researchers who previously have not conducted a large online survey panel.

https://doi.org/10.52202/067871-0006

Figure 1

Factor Structure for the Outpatient EXI

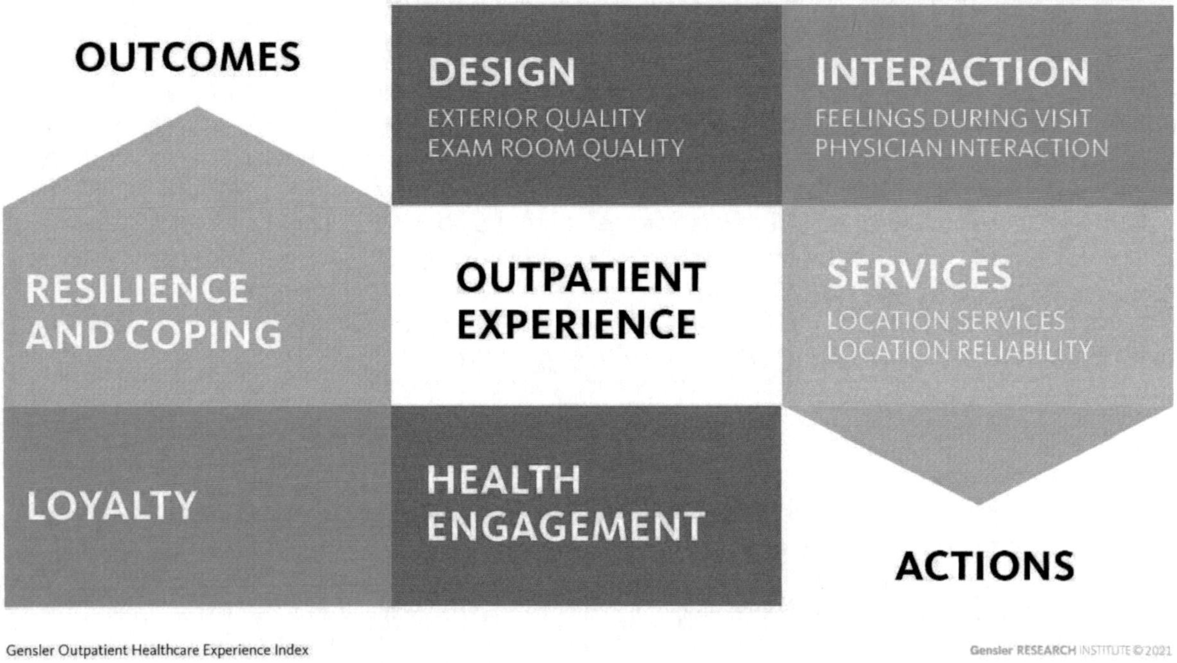

Note. Exploratory and confirmatory factor analyses validated factors that constitute the broader O-EXI Framework and the actionable scores clients and design practitioners use to inform their project goal setting and design decision making. The framework's factors are organized into design and operational "actions" that can shape outpatient consumers' experiences.

Ethics

Larger online panel providers have in-house legal experts. They are invaluable for reviewing questionnaire content for any legal and ethical risks. As a case in point, for the O-EXI, Kantar Profiles' legal experts were promptly advised on questionnaire content for compliance with ISO 20252 and the US Health Insurance Portability and Accountability Act (HIPAA). Though most online panel provider surveys are designed to fall under the heading of market research, healthcare design researchers may want to consider submission of their study to their institution or a reputable third-party institutional review board (IRB) for human subjects review approvals. This additional precaution may appear to be an extra hurdle and unnecessary expense. However, it is worthwhile if the client researcher wants to avoid time-consuming qualifiers for their submissions to conferences and scientific journals that require IRB approval. If the study in fact qualifies as market research and the study is already well planned, then an "Exempt" IRB designation or an expedited IRB review with a short turnaround time would likely be the result. Readers interested in IRB regulations should look at the Federal Code, 45 CFR 46 for the protection of human subjects.

Self-Selection and Satisficing

In general, the team consisting of the online panel provider's survey programmers and administrators and the client researcher need to be on the outlook for panelist self-selection and satisficing (Endres et al., 2022; Zhang et al., 2020). Panelist self-selection occurs when a panelist chooses to opt-in to a panel and take part in a survey or expend less effort when taking a survey. As primary safeguards against self-selection bias, the chosen online panel provider should regularly update their panel with new pan-

 https://doi.org/10.52202/067871-0006

elists by randomly selecting these from the population of interest and implementing the prescreening/vetting quality controls summarized earlier (Sandorf et al., 2022).

The frequent survey responder or "professional survey taker" (PST) has come to epitomize selection bias in online panels. Financial reward, multi-tasking, distractions, and inattention are all thought to influence a PST to satisfice or expend as little thought as possible to meet the bare minimum criteria to complete a survey task. Satisficing, in turn, leads to poor data quality that can negatively impact subsequent statistical assumptions and study goals (Berry et al., 2022; Décieux, 2020).

However, concerns with the quality of PSTs and online panels in general may be exaggerated. PSTs are more likely to be intrinsically motivated to complete surveys and provide higher quality responses, if in a concise manner (Brüggen & Dholakia, 2010; Schwarz et al., 2022; Zhang et al., 2020). Moreover, for panel surveys in general, responses tend to be of higher quality when compared to other self-report methods because of the diversity of panelists, relative anonymity, and less motivation to be socially desirable (Berry et al., 2022).

There are signs of satisficing a client researcher can look out for in their data. Nonsensical responses to open-ended questions are clear indicators of satisficing. "Straightlining" or the same response on succeeding questions and "zig zagging" or "Christmas-treeing" patterns of responses across succeeding questions are strong indicators of satisficing. Nonresponse to items and regular selection of "I don't know" or "Not applicable" options can be concerning. Indicators of speeding and the panelist's inattention are harder to infer from visual inspection of responses.

Survey Development Techniques to Proactively Detect Satisficing

A questionnaire's design and programming are highly effective at proactively detecting both conspicuous and inconspicuous indicators of satisficing. The survey can be programmed to require answers to the majority or all questions. Questions and their responses can be randomized to prevent any order effects. Speed traps can be programmed into the online survey so that panelists who answer a set of more challenging questions or read instructions too quickly can be prompted to reread the instruction or be disqualified. As a precedent, a 300 millisecond per word threshold was used by Zhang et al. (2020). Similarly, if a panelist is distracted from a survey for too long as indicated by a protracted duration of nonresponse, they can be prompted or disqualified.

Reverse-scored questions are another popular way to check for satisficing. For example, in a series of positively-phrased questions asking about a panelist's feelings (i.e., degree of happiness, comfort, or calm on a scale from 1 to 5 with "5" as "Very High"), a client researcher can include questions that are negatively-phrased (i.e., degree of sadness, fear, anxiety, or pain on the same 1 to 5 scale). If responses to the positive and negatively-phrased questions are the same, the responding survey panelist is likely satisficing.

Attention checks are highly effective at detecting satisficing. Ideally, attention checks will be front-loaded in a survey with others randomly scattered throughout the remainder of the survey; a deviant survey panelist who completes most of the survey before being disqualified may argue for compensation for the time and effort taken, regardless of whether their responses were used for subsequent analyses. Also, IRBs differ in whether they require panelists to be paid independent of the extent of completion of the survey.

For the O-EXI, the client researcher asked the panelists to restrict their answers to their last outpatient visit with a physician. "Outpatient visit" and "physician" were defined. Following the definitions, panelists were asked if they read the definitions. A subsequent question then asked the panelists if their outpatient visit to a physician was to a dental care provider, chiropractor, or physician. Surprisingly, several panelists were disqualified from the survey because they claimed their last visit to a chiropractor or dental care provider was an outpatient visit with a physician. A popular attention check is to politely tell the panelist to choose a specific response to a multiple-choice question. Benign bogus or fictitious questions asking panelists if "the sun orbits the earth" or if "Miami is the capital of Canada" with a true or false response are useful. Similarly, a panelist may be asked to enter what year they were born, which allows for

https://doi.org/10.52202/067871-0006

screening of unreasonable answers while also providing useful demographic data. An alternative method asks panelists to answer a very basic arithmetic question and then to type or select the correct numeric value.

Finally, a subset of questions at the end of a survey can query the panelist as to why they took the survey and about their experiences taking this survey, taking other surveys, and with other panel providers. Was the panelist taking the survey in a setting where they may have experienced distractions, been influenced by others, or felt self-conscious? Were they multitasking or distracted while taking the survey? Were they motivated to take the survey for the incentive or another reason? Have they taken several surveys in the past or are members of other panels? Answers to these questions can be considered during analyses (Schwarz et al., 2022; Zhang et al., 2020).

After a survey is closed, it may be too late and costly to detect and replace problematic panelists that are to be excluded from analyses, particularly if there is a high incidence or "screen-out" rate. As an example of how to avoid this situation, a survey programmed in the co-authors' Qualtrics platform is monitored in real-time during fielding with the "Expert Review" feature (Qualtrics, 2022). Infractions are automatically identified, panelists suspended from the survey, and the panel provider can be notified of what panelists need to be replaced. If a platform other than Qualtrics is used by the online panel provider (as was the case for the O-EXI), the co-authors routinely check with the provider on data quality throughout data collection and routinely spot-check the data after receiving a copy of the data and while the study is ongoing.

Sampling and Data Collection: The O-EXI

In November 2019, a draft of the O-EXI was piloted with 1,011 panelists. Because of lessons learned from the pilot, the team decided to let panelists answer questions relative to their last outpatient visit within the last 2 years, instead of the last 12 months. Also, demographic questions were moved to the front of the questionnaire. If a quota had already been filled for a certain characteristic (i.e., female, ethnicity, etc.) then a person presenting to take the survey who had that characteristic would be disqualified from the survey.

A proportional stratified random sampling strategy was used to select panelists from Kantar Profiles' LifePoints panel for participation in the O-EXI (Lohr, 2019). The sample was proportionally balanced to the most recent US Census data on the following strata: gender, region, income, employment, ethnicity, and age. Advantages of proportional stratified random sampling compared to other traditional random sampling include greater accuracy of the survey's values with those of the population of interest and precision or reproducibility of the findings.

With the sudden onset of the COVID-19 pandemic in March 2020, it was necessary to include a subsection to the O-EXI addressing patients' experiences with virtual care. Without a precedent to draw from, we estimated a quota of 500 patients who had used virtual health since the start of the pandemic would be sufficient.

The full study took 5 months to complete from June to October 2020 because of the survey's disqualifiers, indicators of satisficing, and a requirement to answer all questions in the questionnaire. A total of 5,043 panelists were disqualified based on their responses to just three questions. Partway through data collection, Kantar Profiles indicated the 30 -to 40-minute duration to complete the survey could have contributed to a number of panelists choosing not to complete the survey. However, upon further analyses by Kantar Profile's statisticians, these incomplete surveys did not statistically and meaningfully differ from those of panelists who completed the survey.

After data cleaning and preparation, there were 3,578 usable and complete O-EXI survey responses. At the conclusion of data collection, only four characteristics fell at 5% or lower when compared to the US Census: LatinX within 8%; 55 to 70 years of age within 7%; 18 to 24 years of age within 6%. It was decided to include "No income" or "Prefer not to say" in response to the income question. Unfortunately, several panelists chose either option when addressing income. In turn, these selections for income made it difficult to find the difference between the O-EXI sample and US Census target for income. The panelists' ratings were not adjusted or weighted as the panelists were randomly selected and the strata had low sampling error compared to the US Census targets.

 https://doi.org/10.52202/067871-0006

Analyses

The analytical approach to the O-EXI was mostly standard. Factor analyses identified latent variables and demonstrated construct validity. Additional structural equation modeling (SEM) identified indirect and direct relationships among latent variables and individual survey items. Underlying these mostly standard analyses is an assumption of symmetric relationships between variables. A symmetric relationship is one in which an independent variable causes a dependent variable when it is present and similarly the dependent variable is not caused when the independent variable is absent. An illustrative example might be that warm weather is required for tomatoes to ripen, but without warm weather, the tomatoes do not ripen.

Developments in statistical analysis software have made it possible and easier to analyze the same data while allowing for the possibility of asymmetric relationships. Asymmetric relationships are situations where either one of the following scenarios are true:

1. An independent variable's presence will certainly lead to the dependent variable to occur, but if the independent variable is absent, the dependent variable may or may not occur
2. If the independent variable is present, the dependent variable may or may not occur, but if the independent variable is absent, the dependent variable will certainly not occur

Asymmetric relationships measured through fsQCA (fuzzy set qualitative comparative analysis) make it possible to gain a more nuanced understanding of complex interaction effects, even on a singular case basis in small sample size data sets.

Applied to the O-EXI, fsQCA allowed for the examination of interactions between individual characteristics like age, location of residence, distance to physician, education level, and insurance coverage with qualities of their outpatient experience. In turn, the researchers were able to detect different patterns that led to the same outcome of interest, like high patient loyalty. As a tool, fsQCA allows for an understanding of different segments of patients and their experiences so that solutions can be better matched to their specific needs and preferences, even when those segments represent smaller groups. These smaller groups normally would not be detected with analyses assuming symmetrical relationships between variables.

Figure 2

Asymmetrical Relationships Among Variables Found Using fsQCA (Fuzzy Set Qualitative Comparative Analysis)

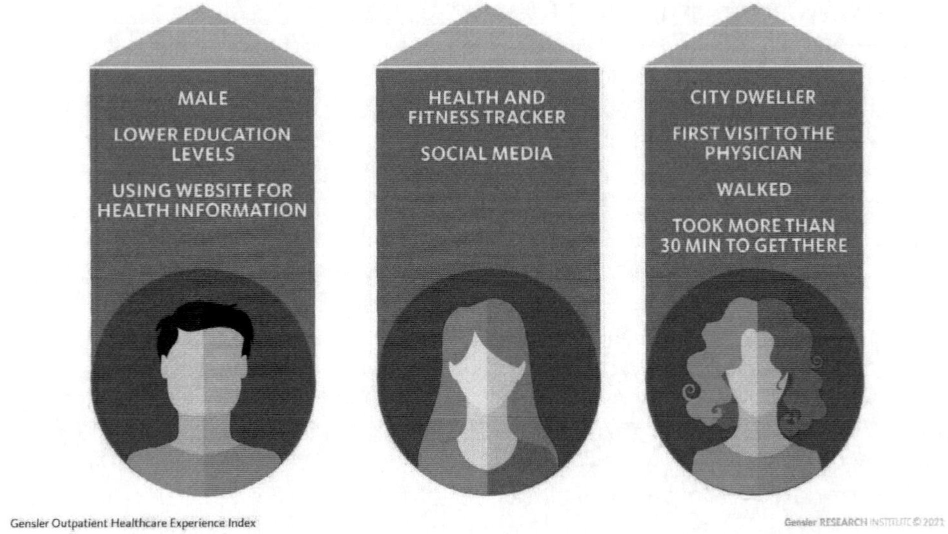

Note. The fsQCAs allowed the researchers to find subgroups of individuals and their characteristics who were influenced to choose one physician over another for their outpatient treatment based on a building's exterior.

Dissemination

Healthcare design research has a legacy of publishing or sharing findings in scientific and refereed venues given the expectations and standards of healthcare designers and their clients. The O-EXI was recognized with a prestigious Environmental Design Research Association Certificate of Research Excellence (EDRA CORE) merit certification in 2022. EDRA CORE was the first and remains the premiere program for blinded and refereed reviews of design research with implications for practice.

However, industry recognition is only one step toward application of research findings. Full translation of research findings needs to occur in those moments of design decision making between a designer and client. With this aim, the O-EXI dashboard is an interactive Tableau-based tool that Gensler's healthcare designers can use and query during work sessions with clients. Similar dashboards are developed for other EXIs and for specific design and research projects.

Acknowledgments

The authors would like to thank Christine Barber, Michelle DeCurtis, June Deng, Dominique Devlin, Isabel Kraut, Minjung Lee, Amira Samiy, and Sofia Song for their contributions to the O-EXI's development and pilot.

References

Brüggen, E., & Dholakia, U. M. (2010). Determinants of participation and response effort in web panel surveys. *Journal of Interactive Marketing, 24*(3), 239-250. https://doi.org/10.1016/j.intmar.2010.04.004

Berry, C., Kees, J., & Burton, S. (2022). Drivers of data quality in advertising research: Differences across MTurk and professional panel samples. *Journal of Advertising, 51*(4), 515-529. https://doi.org/10.1080/00913367.2022.2079026

Décieux, J. P. (2022). Sequential on-device multitasking within online surveys: A data quality and response behavior perspective. Advance online publication. *Sociological Methods Research*. https://doi.org/10.1177/00491241221082593

Devlin, A.S. (2020). *The research experience: Planning, conducting, and reporting research* (2nd ed.). Sage.

Endres, K., Hillygus, D. S., DeBell, M., & Iyengar, S. (2022). A randomized experiment evaluating survey mode effects for video interviewing. *Political Science Research and Methods, 1-16*. https://doi.org/10.1017/psrm.2022.30

European Society for Opinion and Market Research (. (2021). 36 Questions to help commission Neuroscience research. Global Research. Retrieved from: https://ana.esomar.org/documents/36-questions-to-help-commission-neuroscience-research-

Gensler Research Institute. Gensler. Retrieved September 30, 2022, from https://www.gensler.com/gri

International Organization for Standardization. (2019). Occupational health and safety management systems—Requirements with guidance for use (ISO/DIS Standard No. 45001). Retrieved from http://www.iso.org/iso/catalogue_detail?csnumber=63787

Hansen, K. M., & Pedersen, R. T. (2012). Efficiency of different recruitment strategies for web panels. *International Journal of Public Opinion Research, 24*(2), 238-249. https://doi.org/10.1093/ijpor/edr020

Kantar Profiles. (2022). ESOMAR 36 – Kantar's response to ESOMAR's 36 questions. Retrieved from https://www.kantar.com/campaigns/pf/esomar-36

Lohr, S. L. (2021). *Sampling: Design and analysis (3rd ed.).* Chapman and Hall/CRC.

Qualtrics ExpertReview Functionality. (2022, February 17). Qualtrics. Retrieved September 30, 2022, from https://www.qualtrics.com/support/survey-platform/survey-module/survey-checker/quality-iq-functionality/

Robinson, J., Rosenzweig, C., Moss, A. J., & Litman, L. (2019). Tapped out or barely tapped? Recommendations for how to harness the vast and largely unused potential of the Mechanical Turk participant pool. *PloS one, 14*(12), e0226394. https://doi.org/10.1371/journal.pone.0226394

 https://doi.org/10.52202/067871-0006

Sandorf, E. D., Grimsrud, K., & Lindhjem, H. (2022). Ponderous, proficient or professional? Survey experience and smartphone effects in stated preference research. *Environmental and Resource Economics, 81*(4), 807-832. https://doi.org/10.1007/s10640-022-00654-5

Schwarz, H., Revilla, M., & Struminskaya, B. (2022). Do previous survey experience and participating due to an incentive affect response quality? Evidence from the CRONOS panel. *Journal of the Royal Statistical Society Series A, Royal Statistical Society, 185*(3), 981-1003.https://doi.org/10.1111/rssa.12857

Shepley, M. M. (2011). *Health facility evaluation for design professionals*. Asclepion.

Sommer, B. B., & Sommer, R. (1997). A practical guide to behavioral research: Tools and techniques (4th ed.). Oxford University Press.

Zhang, C., Antoun, C., Yan, H. Y., & Conrad, F. G. (2020). Professional panelists in opt-in online panels: What do we really know?. Social Science Computer Review, 38(6), 703-719. https://doi.org/10.1177/0894439319845

Increasing Brand Value by Elevating the Human Experience: Qualitative Considerations

Lorissa MacAllister, *PhD (Enviah)*

ABSTRACT

Increasingly, healthcare providers compete on the strength of their brands for patients, staff, and resources. Since brand value is the product of the aggregate experiences and perceptions of a product's users, it is directly proportionate to the quality of human interactions and encounters with a healthcare provider. To increase the value of its brand, a healthcare provider must satisfy the objective needs and subjective expectations of patients, families, and staff. Focusing on the top line of quality and service results will improve the overall value. With a focus on value, Dr. MacAllister will describe the qualitive data that can be extracted from healthcare organizations (productivity, capacity, efficacy and satisfaction) that directly tie to the outcome of the organizations and how the brand is constructed. She will also share how to apply the data to an operating model focused on designing positive memories and ultimately brand through a deeper application of Dr. Daniel Kahneman's Peak – End rule. Dr. MacAllister's approach assures that there is highest performance in the environment, the most maneuverable elements of the equation to get better results. The limitations of this work focus around the organization's ability to adapt and change to the vision that is established.

INTRODUCTION

As healthcare becomes more commercialized and competitive, there is the onset of for-profit retailers, from Amazon to Walmart, entering the marketplace for the delivery of healthcare. Healthcare is embarking on a transformation of how to capture the value and create a branded experience that retains and expands its business in a way that has never been seen before within the healthcare domain. This paper explores the way a healthcare system can use evidence, both qualitative and quantitative, to improve its overall brand value with the goal of elevating the human experience.

What is Value?

Value and healthcare have been elusive terms that emerged as primary drivers to high-quality healthcare with the onset of value-based purchasing. Prompted by Porter and Teisberg's (2006) initial nudge to shift the system of evaluation, the US Healthcare system, including Medicare and Medicare, shared services programs, and other providers, have spent the last 20 years focused on defining and bringing clarity to healthcare value. Explicit clarity regarding the meaning of value for the field has not yet been agreed upon, as noted in the Landon et al. (2021) literature review. What is clear is that we can use a framework for assessing value in three categories: Components, scope, and perspective (see Figure 1). In many discussions, the role of perspective is missing as a key variable. This exploratory paper intends to further solidify the role of perspective in value achievement; without perspective, one can never achieve brand value and a deeper engagement with clients.

 https://doi.org/10.52202/067871-0007

Figure 1
Value Categories and Their Elements

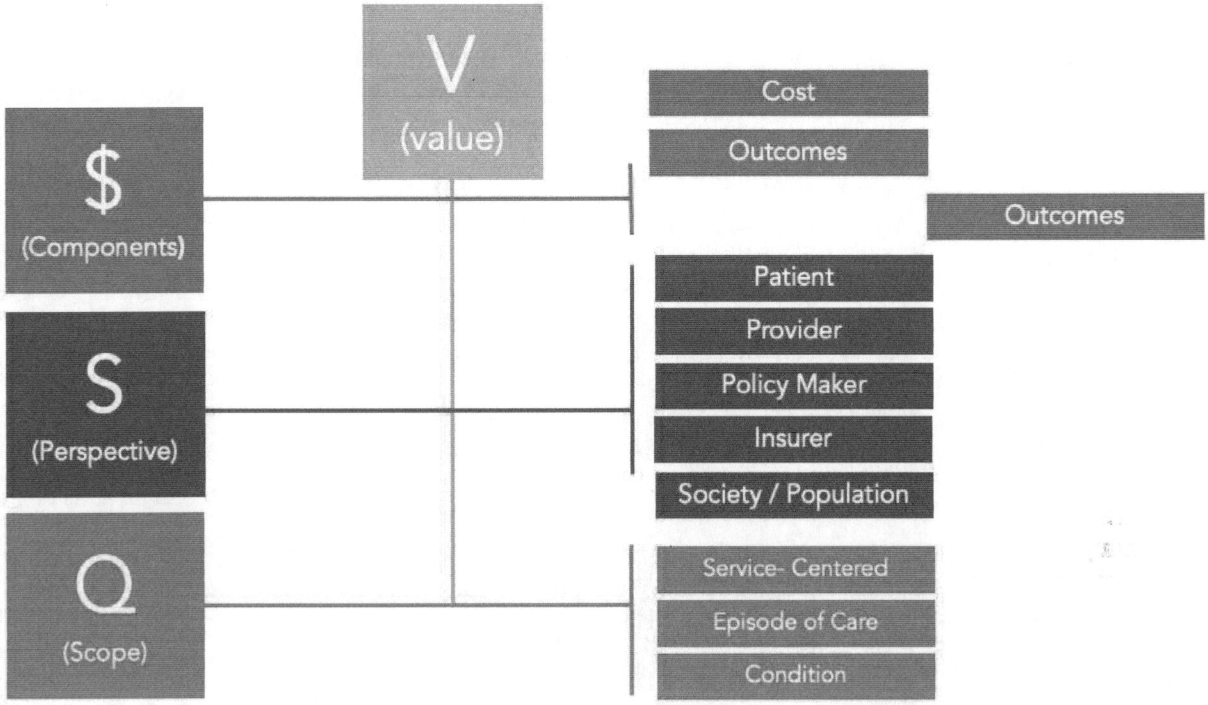

Figure 2
Value Equation

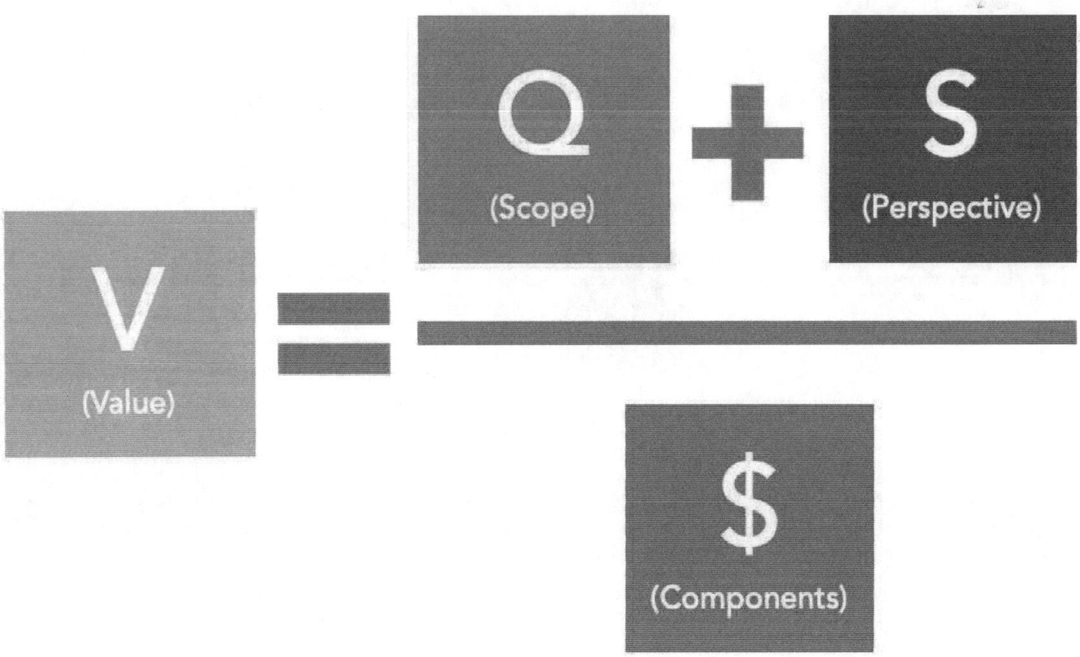

https://doi.org/10.52202/067871-0007

In summary, the components are the anchor of the qualitative outcomes and are primarily defined as cost and outcomes such as quality harms and benefits. These themes and elements have been explored and are the focus of exploration in the field (Landon et al., 2022). An example of these components that define value appears in Owens et al.'s (2011) article that explored value in three concepts: benefits& harms, cost including downstream cost, and incremental cost-effectiveness ratio.

The scope of value determines the area of focus such as a condition, service, or episode of care. The scope narrows the field of value and determines the population and possible areas of influence.

The most elusive and complex factor is the perspective of value. Within healthcare, perspective is one of the most complex for all involved: patients, providers, policymakers, insurers, and the greater society at large. The varying perspectives of these constituencies can shift and significantly change the outcome of value.

With the categories and the established framework of value understood, we now need to build an application showing how to achieve value. To do so we look to the Institute of Medicine's (2001) definition of value as quality over cost. A more in-depth framework using the formula put perspective and scope as the numerator and components as the denominator. Now it is up to each entity to adjust the customized formula to achieve clear value for their entity; thus, the equations change and evolve depending on the focus. Landon et al. (2021) shared that the many elements of value are used commonly in healthcare evaluation, however, the final unified definition is not universally adopted. It is customized to the organization and system. The primary importance is that the formula contains both a scope and a perspective over a component. While there will likely not be one universally accepted definition for healthcare quality, the equation is clear. To achieve value for any organization this understanding will need to be defined.

Now that value and the elements are understood and the equation is defined, how can an organization build value with its brand? The perspective of the person is the key element to strengthening both value and brand. To demonstrate this connection, we now dive deeper into the brand and how our brain creates our perceptions.

Figure 3

Illustrations of High-Value Brands

Maurya et al. (2012) identify a Brand as a symbol, sign, or design that identifies a good or service. Therefore, the Brand can be designed for the experience of healthcare. Brand value emerges when you attach a meaning to a symbol that is positive and attractive (see Figure 3). One goal of brand building is to influence the stories that are being told about you and your organization. Building a brand can be done in many ways, but first, we turn to the science of storytelling and how we experience creating and crafting stories.

To understand the science of storytelling we turn to the Nobel prize-winning economist Daniel Kahneman, whose psychological research has studied the science of design making and how the brain

 https://doi.org/10.52202/067871-0007

has both a fast and a slow part that continually challenges us. Kahneman (2011) shared that there are "Two Selves;" there is the fast-experiencing self that is in the moment and the "now;" the fast linear self is unconscious, automatic, and quick and its memories last only seconds and vanish (see Figure 4).

The remembering self is more rational and slower with heightened self-awareness and deliberately focused on rational thinking logic as it seeks new and missing information to make decisions (see Figure 4). Kahneman's (2011) research found that most of our thinking happens in the fast or experiencing self as we repeat tasks more on autopilot so we can rapidly move to the next thing more automatically and heuristically.

Figure 4
The Two Selves: Experiencing & Remembering Self

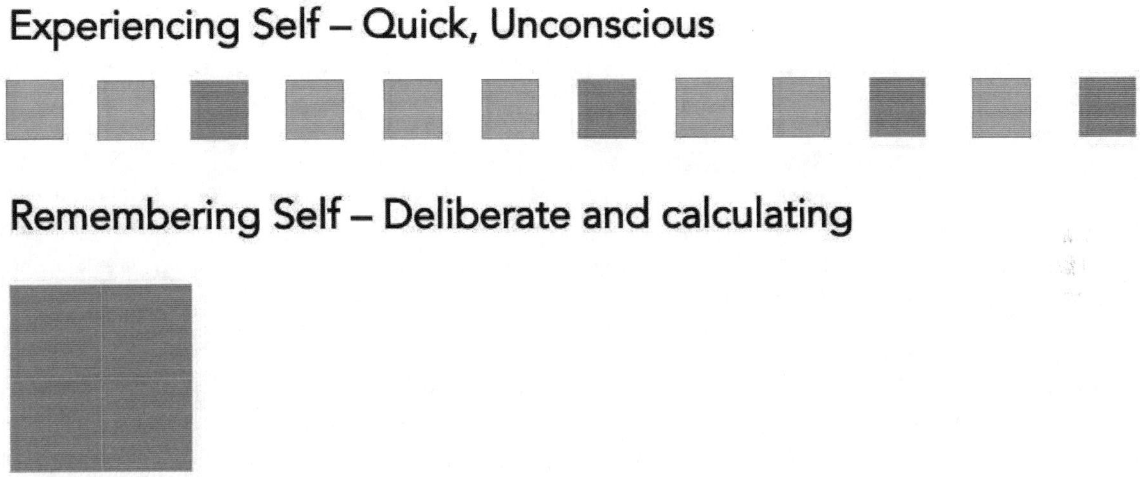

The slow-processing self is more reflective as it remembers the moments of change or the ending of the automatic experiencing self. The remembering selves collect those memories our mind perpetually processes memories and organizes and interprets continually producing your personalized cognitive program. Each person has a unique cognitive program that is shaped by their life and experiences and becomes their database for their decisions and perceptions of future decisions (see Figure 5). We tell stories to make sense of our cognitive program selecting portions of our memories and "filling" the in-between with familiar concepts that bring meaning and identity to the memories (see Figure 5).

Figure 5
Perceptual Processing

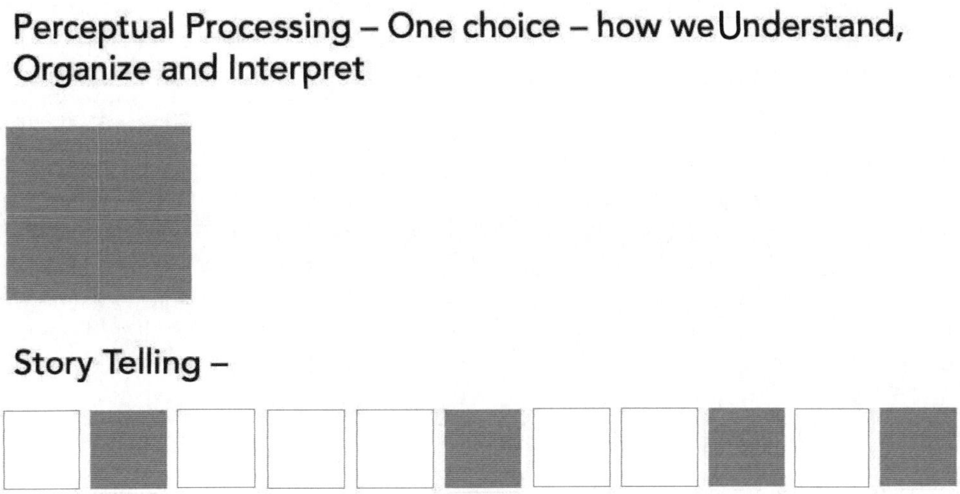

https://doi.org/10.52202/067871-0007

Therefore, if one desires to influence a story that becomes a brand that is being told about them the memories retained about that brand story will need to be further developed. As Kahneman (2011) discovered, the memories that people retain are not the sum of everything that occurred but rather the changes or peaks and the ends in the automatic experiencing self, called the peak-end rule (Redelmeier & Kahneman, 1996, p. 380).

With that understanding and application in healthcare, one can look at the patient experience and identify what situations would influence a change in the experience such as pushing a call button for assistance when you are in your room or moving to another service within the building to finally discharge. In other words, what situations exist in the healthcare system to shape a patient's perception of the brand? The brand is the story that is told about an organization, and if we know how the brain works and processes experience one can focus on the critical points in the delivery of care that truly influence the brand and ultimately the value of care.

Improving Healthcare Experiences

Recognizing that perception is a connective element that can build stronger value and brand, how can one create or design a system that improves not only the perception of care for the "self" but also the actual care experience? The environment in many instances has been missing in the understanding of how one's experience as a patient could be influenced. Recent research by MacAllister et al. (2019) demonstrates that the environment can directly influence the experience outcomes and further enhance the brand and value of the organization. Let's now use the MacAllister et al. study as a direct application of how the environment can influence brand value.

Brand – Experiencing Self

MacAllister et al. (2019) explored multiple inpatient nursing units at Emory University Hospital and statistically found a relationship between the room and unit layout and the perception of care. In this work, we will look at one outcome situation that can directly tie to a change in the experiencing self, the moment someone enters your inpatient room. This measure is defined as the location of the first encounter. Again, with an understanding of Kahneman's (2011) two selves the first encounter records the change in an experiencing self when a staff member enters the room. The first encounter environmental measure is the orientation of the hand wash sink and its relationship to the view of the patient's head. This measure of the first encounter can be determined based on the requirement in the United States healthcare system for any staff member entering the room to wash their hands. When a staff member enters an inpatient room, it creates a change of experience for the patient in that room.

Brand – Remembering Self

For the second self, the remembering self, we turn to the patient experience scores collected by the hospital. These scores are collected and recorded in the room where the patient was staying from which the patient was discharged. The patient experience survey questions range from the environment to encounters with various staff members and services that occurred during their stay at the hospital and are administered by a third-party service. This survey serves to record the remembering self in a specific context of a hospital stay and experience. MacAllister et al. (2019) found that there was a statistically significant relationship with the point of the first encounter with sinks that were positioned to see the head of the patient within a 30-degree purview. In Figure 6 the three questions that had higher top box scores (measuring only the highest score) were in the following areas: Skill of the physician (5% higher), Staff concern for privacy and friendliness (25.2% higher), and Courtesy of the nurse (21.8% higher). These findings could plausibly demonstrate that the patient's remembering was self-determined that when the nurse entered the room and was able to make eye contact with the patient while washing their hands, the patient perceived the nurse as friendlier and with a deeper concern for the patient's needs. In addition, the skill of the physician was also higher in rooms with a hand wash sink oriented toward the head of the patient's bed. This study further quantifies the role the environment plays in improving the brand story of a hospital. Now let us explore how this same study could be applied to improving the value of that hospital.

 https://doi.org/10.52202/067871-0007

Figure 6

The Science of Remembering Self: Application of Practice to an Inpatient Hospital Room

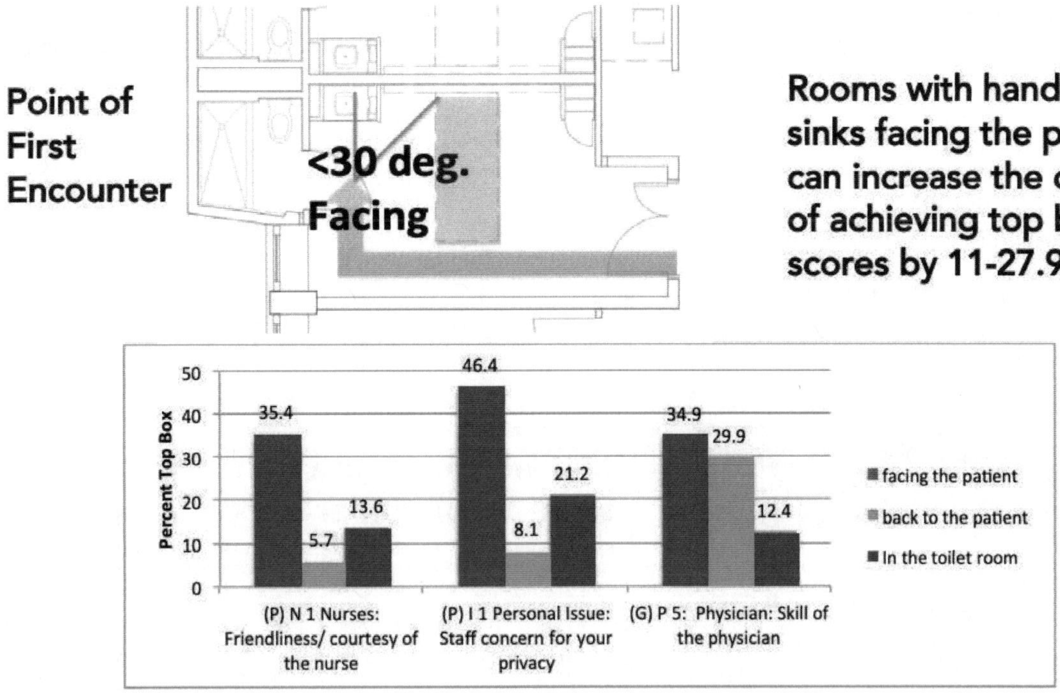

Value – Equation

The value from this study can be defined by the value of the room performance. The value equation noted in Figure 7 clearly defines that the scope is the staff encounter with the patient and the perception is the perspective of care of the patient over the experience outcomes. The improved value is the overall improvement in satisfaction score that directly improves the revenue by at least 1% in revenue for each percentage improvement (Buhlman & Lee, 2019). This relationship shows a clear improved direct value to the hospital system and further confirms the pivotal role the environment can play in improving overall brand value to the organization.

Figure 7

Components of Value for Sensory Experience

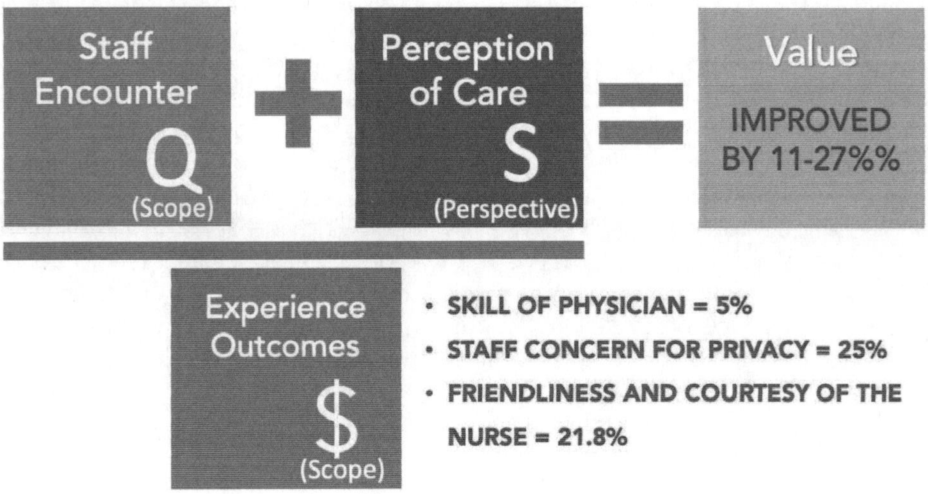

https://doi.org/10.52202/067871-0007

Together, the improved brand and the value elements directly can create standards and practices focused on health to achieve the highest measurable outcomes that bring value to our hospital clients and their patients!

Designing High-Value Brand

Given the research and application of both brand and value, how can practitioners create and design high-value brands? The focus here is again the work of Kahneman (2011); he shared in his research the importance of experiencing self-leading over 90% of the decisions; the remembering self is framed by the changes in the experiences or the end. This framework Kahneman defined as the Peak and End rule. As designers, we can design high-value brands if we follow three easy steps: Frame, Define, and Interpret (see Figure 8).

Figure 8

Designing High-Value Experiences

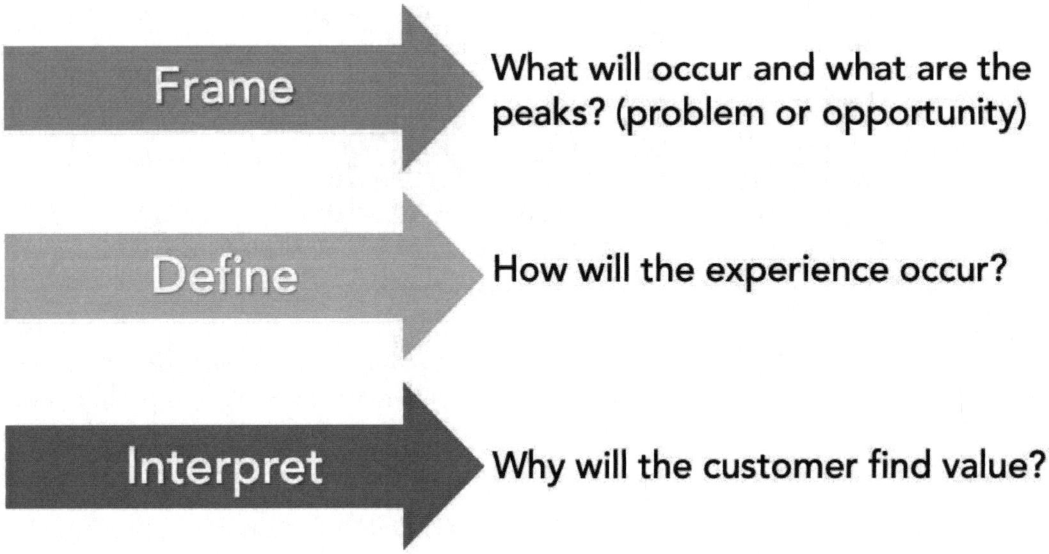

First, if we know that the changes are central to framing the remembering self, we need to take the journey of the patient and identify the peaks in the experience of care. We can define the peaks as "critical touch points." These touch points are the critical points of change in the journey and the patient will likely remember and frame the experience from that change -- either positive or negative.

The next step is to define more clearly what is occurring at the peak. Take the time to clearly state the experience that is desired and the behaviors that the staff would exhibit to achieve the experience, and then most importantly document the research evidence showing why or how this outcome could be derived. Like the determination of the point of the first encounter, it has been demonstrated through research that the environment directly influences the satisfaction outcomes of certain questions. This understanding and knowledge of the framework are key to applying and creating value in practice.

Finally, the application to practice needs to be interpreted and predicted to determine the outcome of overall value to the situation. The improved value is defined by the equation of scope and perspective over components. This practice with a focus on improving the outcome will provide the needed changes in healthcare practice as many organizations miss the opportunity to look at the whole system and approach to gain a higher outcome. It takes both a clear brand experience and a focus on the key elements of value to truly achieve the improvements that the system desires.

 https://doi.org/10.52202/067871-0007

References

Buhlman, N. W., & Lee, T. H. (2019, May 08) When patient experience and employee engagement both improve, hospitals' ratings and profits climb. *Harvard Business Review*. https://hbr.org/2019/05/when-patient-experience-and-employee-engagement-both-improve-hospitals-ratings-and-profits-climb

Institute of Medicine (US) Committee on quality of healthcare in America. (2001). Crossing the quality chasm: A new health system for the 21st century. *National Academies Press (US).*

Kahneman, D. (2011). Thinking, fast and slow. *Macmillan.*

Landon, S. N., Padikkala, J., & Horwitz, L. I. (2021). Defining value in health care: A scoping review of the literature. *International Journal for Quality in Health Care, 33*(4). https://doi.org/10.1093/intqhc/mzab140

MacAllister, L., Zimring, C., & Ryherd, E. (2019). Exploring the relationships between patient room layout and patient satisfaction. *HERD: Health Environments Research & Design Journal, 12*(1), 91-107. https://doi.org/ 10.1177/1937586718782163.

Maurya, U. K., & Mishra, P. (2012). What is a brand? A perspective on brand meaning. *European Journal of Business and Management, 4*(3), 122-133.

Owens, D. K., Qaseem, A., Chou, R., Shekelle, P., & Clinical guidelines committee of the American College of Physicians (2011). High-value, cost-conscious health care: Concepts for clinicians to evaluate the benefits, harms, and costs of medical interventions. *Annals of Internal Medicine, 154*(3), 174–180. https://doi.org/10.7326/0003-4819-154-3-201102010-00007

Porter, M. T., & Teisberg, E. O. (2006). *Redefining Health Care: Creating value-based competition on results*. Havard Business School Press.

Redelmeier, D. A., & Kahneman, D. (1996). Patients' memories of painful medical treatments: Real-time and retrospective evaluations of two minimally invasive procedures. *Pain, 66*(1), 3–8. https://doi.org/10.1016/0304-3959(96)02994-6

Wood, L. (2000). Brands and brand equity: Definition and management. *Management Decision, 38*(9), 662-669. https://doi.org/10.1108/00251740010379100

https://doi.org/10.52202/067871-0007

Packaging for the Practitioner: The Role of Secondary Research

***Ellen Taylor**, PhD, AIA, MBA, EDAC (The Center for Health Design)*

ABSTRACT

We live in an era of information overload or information pollution. Two decades ago, Jakob Nielsen stated that the web was a procrastination apparatus that "can absorb as much time as is required to ensure that you won't get any real work done." Today, access to the Internet can be a blessing and a curse for practitioners who can Google almost anything, and for some, this constitutes a form of research. The reality is that just like all other types of "searchable" information, rigorous research is being published at a rate that no one can realistically follow. As a result, the value of secondary research is increasingly important. This is especially true for practitioners in environmental design where the time and resources to access peer-reviewed research may be more challenging than for their counterparts in academia.

The dissemination of such secondary research can come in a variety of forms, and with a wide range of outputs, the process for creating tools and resources can also vary in rigor. The Center for Health Design has a history of summarizing relevant evidence for design practitioners who focus on the influence of the built environment on a host of outcomes. These tools and resources include peer reviewed publications, Post Occupancy Evaluations, Safety Risk Assessment tools, Interactive Diagrams, Slidecasts, Issue Briefs, etc. As part of this session, participants will learn how packaging for the practitioner involves the creation of its own study design with a choice of research methods (primary and secondary) that need to appropriately address the aims of the deliverable, while at the same time taking into account the age-old triad of scope, schedule, and budget.

INTRODUCTION

Many practitioners do their best to stay abreast of what we know from the latest research. The reality is that just like all other types of "searchable" information, rigorous research is being published at a rate that no one can realistically follow. It is like drinking from the infamous firehose. As a result, the value of secondary research is increasingly important and especially the case for practitioners in environmental design where the time and effort to access peer-reviewed research may be more daunting than for their counterparts in academia.

As a researcher thinking about how to frame a discussion about secondary research, I naturally started with a search of the literature. It seemed that there would be a body of applicable knowledge to use as a starting point. Imagine my excitement when one of the first papers to come up from my search terms was "Recovery of gold from secondary sources – A review." I thought this document would surely be my starting point until I found this was a scholarly paper that addressed the recovery of gold (the metal) from spent sources through mechanical separation methods such as bio-hydrometallurgical technologies (Syed, 2012). While this outcome was not exactly what I had wanted, the example illustrates the challenges we all face when searching for information –an endless cycle of suggested searches and

 https://doi.org/10.52202/067871-0008

diversions before you realize that you have spent an hour or more becoming more "educated" but with no answer in hand.

This paper focuses on five challenges practitioners face in finding research and other forms of evidence to support decision-making. These challenges include the availability of time, searcher preferences, sources of dissemination, communication methods, and the quest for specific solutions. Each challenge is followed by examples of how The Center for Health Design (The Center) uses secondary research to respond to industry realities.

Challenge 1: Time Searching

> *"The Web is a procrastination apparatus: It can absorb as much time as is required to ensure that you won't get any real work done"* (Nielsen, 2003, online section 3).

Two decades ago, usability advocate Jakob Nielsen opined about Internet pollution with the web as a procrastination apparatus. At that time, between 608-682 million people worldwide used the Internet (Internet Growth Statistics 1995 to 2022 - the Global Village Online, 2022), and there were fewer than 41 million websites (Total Number of Websites - Internet Live Stats, 2022). In mid-2022, there were 5.3 billion Internet users with more than 1.9 billion websites (Internet Live Stats - Internet Usage & Social Media Statistics, 2022). Internet Live Stats (2022) also indicated that in less than one day, more than 200 billion emails had been sent, Google was used to search more than 6.2 billion times (more than 100,000 searches per second), and there were 151,408 Gigabytes on Internet traffic every second.

It is not hard to see how Internet access can be both a blessing and a curse for practitioners who can Google nearly anything, and for some, this constitutes a form of research. However, it is not just information overload – it is the hours people spend trying to find what they are looking for, often following the endless rabbit holes of suggested terms offered by search engines.

In a world where your design was needed yesterday, few practitioners feel they have the time to dedicate to reading research. A study conducted by educator Ann Huber (2016b) investigated how interior designers used research in decision-making and found that nearly half (49%) spent less than 10 minutes reviewing a source. Many indicated that time was a constraint in reviewing research (see Figure 1). Seek, and you might find what you are looking for, but you may just be wasting your time.

Figure 1

Time Spent Reviewing Research/Barriers to Using Research

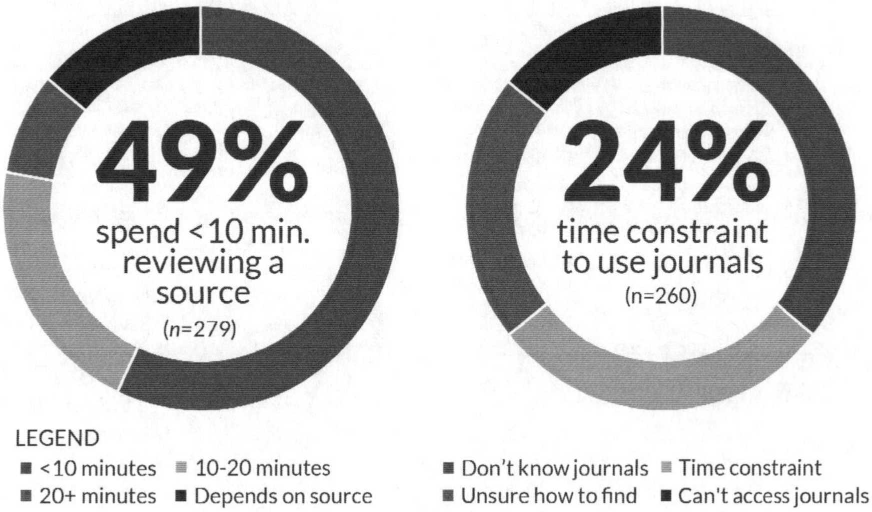

Note. The data are from "Research utilization in the design decision making process," by Amy Huber, 2016, International Journal of Architectural Research: ArchNet-IJAR, Volume 10 (Issue 1), p. 14, 16. https://doi.org/10.26687/archnet-ijar.v10i1.824. Copyright 2016 by Archnet-IJAR.

Opportunity 1: Less Time Searching through Evidence Summaries

When trying to find information on a topic and establish the base of knowledge for research in the built environment, a literature review is often a starting point. The rigor of literature reviews has evolved in the past 20 years, and the industry has increasingly shifted from traditional and narrative reviews to peer-reviewed structured systematic and scoping reviews. A systematic review is typically used to answer a specific research question, while a scoping review typically offers a more breadth than depth, in order to summarize the overall state of knowledge on a less-studies topic (Arksey & O'Malley, 2005). There is a time and place for different types of reviews as each takes varying levels of effort to create.

The Center produces narrative reviews in the form of Issue Briefs on specific topics that are important to the industry (e.g., infection control, clinician burnout), but The Center's research team also publishes in peer-reviewed journals when there is a specific identified industry need. For example, the Facility Guidelines Institute (FGI) was interested in understanding how the evidence had evolved in the literature since the 2006 decision to require private patient rooms in newly constructed US hospitals. The result was a systematic review evaluating a decade (2006-2016) of published outcomes associated with single-occupancy patient rooms (Taylor et al., 2018). As a systematic review is intentionally focused, it can be a time-saving solution to evaluate outcomes for specific design decisions that are being considered. Given that designers may only spend 10 minutes reviewing a source, it can also be important to visually summarize the information through figures and tables (see Figure 2).

Figure 2

Partial Summary of Literature Review Findings

Broad Outcome	Outcome detail	First Author, Year	Results	1	2	3	4	Country	Setting
Communication	Call Response	Deitrick, 2010	+	●				USA	M-S
	Type/Duration	van de Glind, 2008	+/o		●			NLD	Urol
Falls		Singh, 2015	-	●				UK	Med/Eld
		Knight, 2016	o	●				UK	Dem
		Maben, 2015	o		●			UK	Mult
ICU delirium		Zaal, 2013	+/o			●		NLD	ICU
		Caruso, 2014	+/o				●	BRA	ICU
Restraints (e.g., rails)		Krüger, 2013	-			●		GER	Hosp
Sleep quality (noise)		Park, 2014	o		●			KOR	Med
User opinion: End user	6	Ward Casscells, 2009	+	●				USA	MHS
User opinion: (multiple)	1, 3, 4, 5, 6, 7, 8, 9, 10, 11	Maben, 2015	+/-		●			UK	Mult
	1, 2, 4, 5, 6, 7, 8, 10	Williams, 2015	+/-			●		UK	Pall
	5	McKeown, 2015	+				●	IRE	Pall
	1, 2, 4, 7, 8, 9, 10, 11, 12	Persson, 2012	+/-				●	SWE	Sur/Or
	1, 3, 5, 7, 10, 12	Ferri, 2015	+/-				●	CAN	ICU

LEGEND

1: Communication
2: Concern for others
3: Isolation from peers
4: Noise/sleep
5: Patient Care
6: Preference
7: Privacy, Companionship, Isolation
8: Personal space
9: Roles in Care
10: Safety
11: Staff views of patient perceptions
12: Workspace

MMAT: 4-pt. scale (4 high) (Pluye et al. 2009)

■ Non-randomized
■ Descriptive
■ Qualitative
■ RCT
■ Mixed Methods
■ Included in narrative synthesis

+ Advantages only
+/o Advantage/neutral
+/- Advantage/disadvantage
o Neutral/no difference
- Disadvantages only

AUS: Australia
BRA: Brazil
CAN: Canada
EGY: Egypt
EU: Europe
FRA: France
UK: United Kingdom
GER: Germany
HKG: Hong Kong
IRE: Irleland
KOR: Korea
NLD: Netherlands
POL: Poland
SWE: Sweden
TUR: Turkey
USA: United States

BMT: Blood/bone marrow transplant
Dem: Dementia
Eld/Reh: Elder/Rehab
Hosp: All hospital
ICU: Intensive Care
Onc: Oncology
PSY: Locked Psych
MHS: Military Health System
M-S: Med-Surg
Med: Medical
Surg: Surgical
Surg/Or: Surgical/Orthopedic
Pall: Palliative
Urol: Urology
Mult: Multiple units

Note. Adapted from "Single-Occupancy Patient Rooms: A Systematic Review of the Literature Since 2006," by Ellen Taylor, Alan J. Card, and Melissa Piatkowski, 2018, HERD: Health Environments Research & Design Journal, Volume 11, Issue 1, p. 89. https://doi.org/10.1177/1937586718755110. Fair use by the authors.

 https://doi.org/10.52202/067871-0008

Challenge 2: Information Seekers

> *Our present information-rich society demands increasingly more of our capacity to handle information, both in terms of quantity and quality. ... The ways each information seeker meets these demands and chooses to handle information-seeking tasks are, however, varied and somewhat unpredictable.* (Heinström, 2005, p. 228)

In a study of personality and study approaches of 305 Finnish Master's thesis students, three information seeking patterns emerged from the factor analysis: fast surfing, broad scanning, and deep diving (Heinström, 2005). Table 1 outlines the characteristics of each information seeker type found in Heinström's study.

Table 1
Summary of Information-Seeker Characteristics

Fast surfers	Broad Scanners	Deep Diver
Gathered information to complete the task, with little attention to content depth.	Characterized by curiosity and happenstance (accidental information discovery).	Considerable effort put into information seeking for highest quality content.
• Spent little time seeking information • Valued the appearance of the document • Preferred an overview of the material with easy-to-understand language. • Felt pressured by time • Challenged in judging relevance	• Wide and thorough searches that are unstructured and develop gradually • Ability to critically evaluate information • Results with slightly related documents through multiple sources (online journals and other material; media; companies, conferences, and associations; books and brochures)	• Valued quality versus quantity • Advance precise search planning • Recognized and reputable authors and sources • High scientific quality • Thoroughness of materials

Note. Summary from "Fast surfing, broad scanning and deep diving: The influence of personality and study approach on students' information-seeking behavior," by Jannica Heinström, 2005, Journal of Documentation, Volume 61, Issue 2, pp. 228–247. https://doi.org/10.1108/00220410510585205.

Opportunity 2: Consolidating Evidence Sources for Information Seekers

The Center for Health Design's website provides a comprehensive approach to compiling evidence in a variety of approaches tailored to different types of information seekers. In addition to the more traditional drop-down menus, the three most accessed resources (Interactive Design Diagrams, Safety Risk Assessment, and Knowledge Repository) are easily accessed from large identifiable links on the homepage (see Figure 3).

Figure 3
Screenshot of The Center for Health Design Homepage

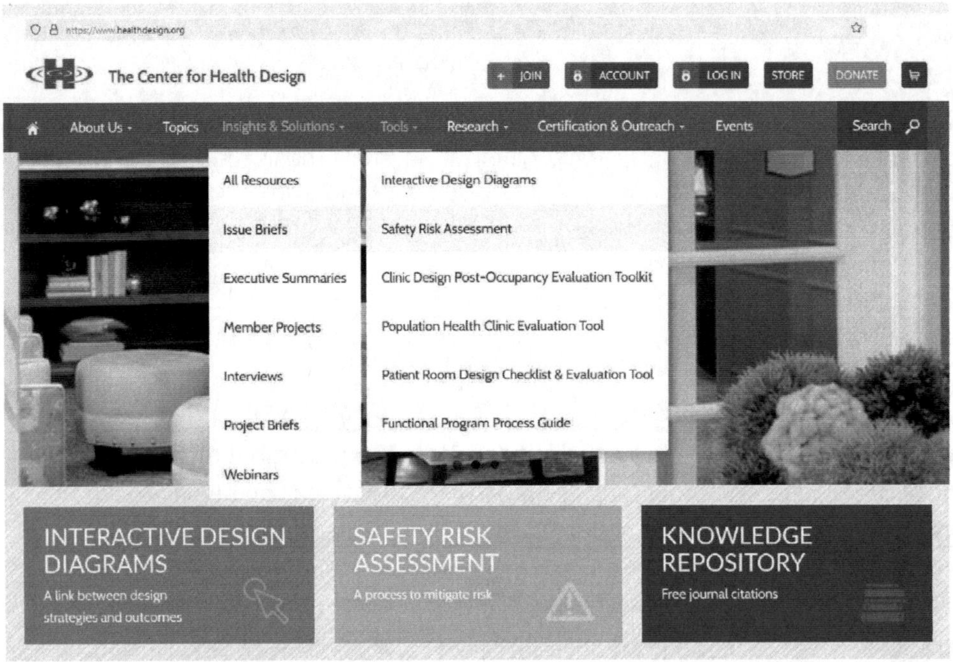

Surfers may check the interactive design diagrams (see Opportunity 4 – Communication with a Visual Representation of Research), which provide a visual overview of the research, while scanners may choose to look at all of the topics and related resources that might be of interest. Deep divers may choose to use the many research-based tools and Knowledge Repository as a way to ascertain the most credible information (see Opportunity 3 – Finding the Dissemination of Research and Opportunity 5 – Creating Answers via Tools for Decision-Making).

Challenge 3: Dissemination

> *"Research findings that remain unknown to design practitioners stand little chance of improving their design processes or project outcomes"* (Huber, 2018, p. 1)

Many researchers strive to publish in peer-reviewed journals. Unfortunately, many designers are unsure of what journals to search, they are unsure how to find the information, and they often do not have access to the papers they seek (Huber, 2016b). Further, Huber (2016) states that only 12% of interior designers used academic journals, relying more often on product literature or conference presentations (see Figure 4).

https://doi.org/10.52202/067871-0008

Figure 4

Sources of Information Used by Interior Designers

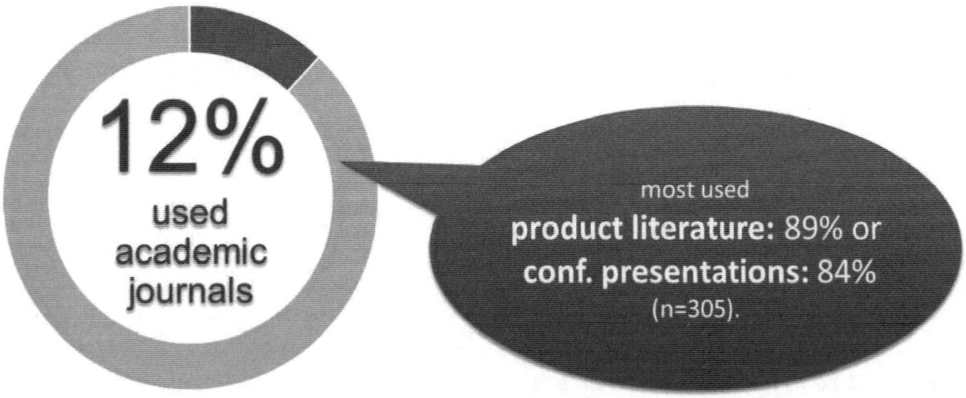

Note. The data are from "Research utilization in the design decision making process," by Amy Huber, 2016, International Journal of Architectural Research: ArchNet-IJAR, Volume 10 (Issue 1), p. 14. https://doi.org/10.26687/archnet-ijar.v10i1.824. Copyright 2016 by Archnet-IJAR.

Opportunity 3: Finding the Dissemination of Research

In 2012, The Center for Health Design launched the Knowledge Repository. Developed with financial support from several industry partners (American Institute of Architects Academy of Architecture for Health [AIA AAH], FGI, American Society of Healthcare Engineers [ASHE]). The Repository is a complete, user-friendly database of healthcare design citations that continues to grow with the latest research.

Every 2 months, the added citations are recorded in an email alert, "Research in a Snap." The database has grown to more than 5,000 citations and includes more than 800 key point summaries, a layperson's narrative of the paper that includes the objectives, methods, findings, and design implications. Studies can also be summarized in a 5-minute (or less) video format (slidecasts or ledes). The search function allows a quick way to find information about a topic, which can be filtered by year of publication or paper type (e.g., literature review). The source link redirects to the publisher's site. While access to the Knowledge Repository is open to anyone, those who create an account can log in and bookmark citations of interest.

For a deeper dive into a specific paper, users can view the EBD (evidence-based design) Journal Club webinars. This webinar is a format where attendees are provided a copy of a recent published study, and on the day of the event, the authors provide a brief summary of their study followed by an extensive question and answer session where attendees can ask about the paper, including information that may not even be included in the publication.

Challenge 4: Communication

> *"Participants in innovation [EBD] are heterophilus. The change agent [researcher] is more technically competent [has training in research]. This often results in ineffective communication due to language differences [research vs design],"*(Bracketed language is an adaptation of from Rogers, 2003, p. 19)

In a study that investigated the visual persuasion of research with design practitioners (n = 34), Huber found 48% of practitioners said they would be more likely to read the graphic-intense document, whereas only 22% indicated the same for a narrative-intense document (Huber, 2016a). The graphic format offered more of a visual roadmap that allowed the reader to skip portions deemed less relevant; while this approach may be preferred by the partitioner, it is important for researchers who may be concerned about the "cherry-picking" of information out of context, to better understand how to best relay content and findings. Huber also found that while images were deemed important, participants felt images should not be gratuitous and only be used to directly convey information or help with content navigation.

Opportunity 4: Communication with a Visual Representation of Research

The Center for Health Design has annually produced interactive and visually engaging annotated diagrams to link design features to desired outcomes. To improve the usability of this growing set of Interactive Design Diagrams launched in 2014, these rooms are categorized in one three different types of healthcare interior environments (inpatient care, outpatient/ambulatory care, home environments).

Each diagram includes "dots" to mark relevant design features (e.g., walls, floors). The user can move the cursor across each dot (e.g., patient handling equipment) to access specific features (see Figure 5).

Figure 5

Interactive Design Diagrams Screenshot (Design Features)

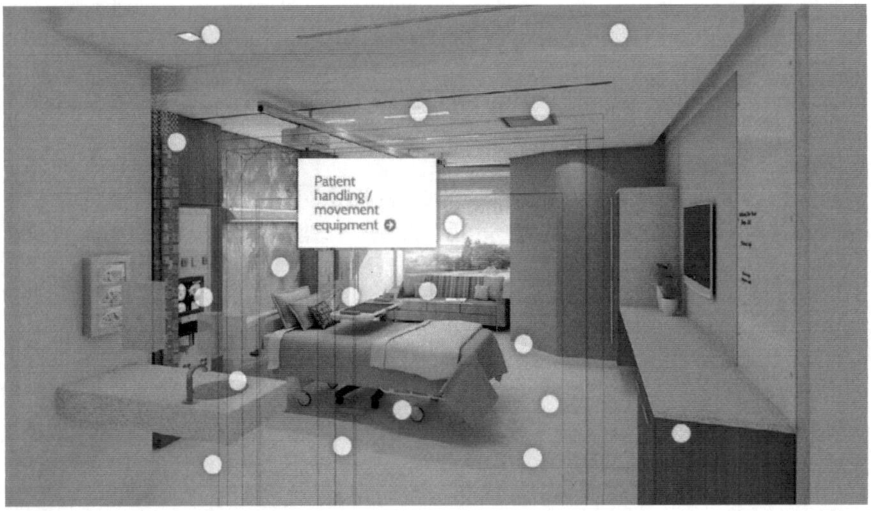

Figure 6

Interactive Design Diagrams Screenshot of Design Features (Desired Outcomes and Strategies)

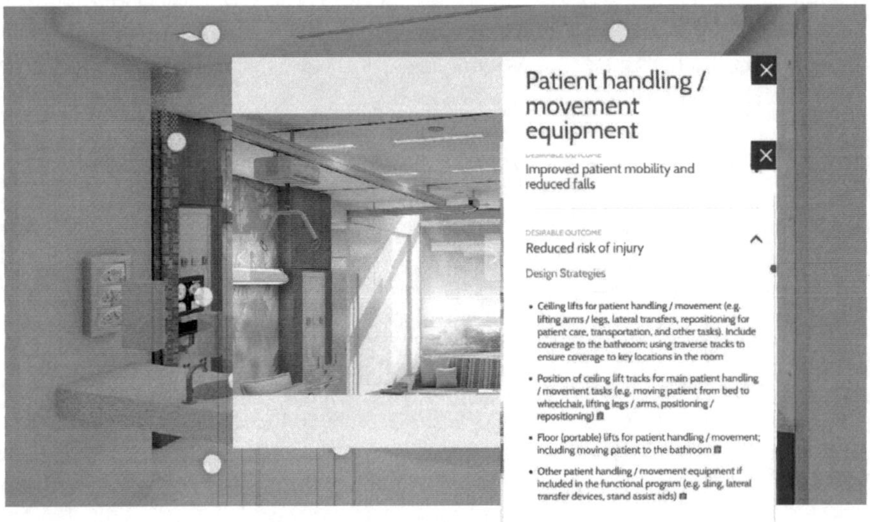

 https://doi.org/10.52202/067871-0008

When design features are selected, the associated desirable outcomes are listed, along with the design strategies that optimize specific outcomes (see Figure 6). Relevant research citations are indicated (where available for the strategy). As a quick check, a 'surfer" can advance through each of the desired outcomes and ask, "Have I considered this?"

Challenge 5: Seeking 'Answers'

> *With respect to the issue of communicating scientific uncertainties, it seems that interpretations of these are likely to be guided by what people think the purpose of science is—specifically, whether science is seen as the quest for absolute truth or as a debate between alternative positions.* (Rabinovich & Morton, 2012, p. 993)

In healthcare design, practitioners often desire specific solutions – absolute truths that are not confounded by complexity. Of course, there is rarely such a simple solution, as each decision should be evaluated in the context of the patient and staff demographics, the model of care, and the context of the setting. Those less familiar with conducting and using research may have less appreciation for a continual evolution of what we do and do not know.

Opportunity 5: Creating Answers via Tools for Decision-Making

The Center uses an approach of secondary research to develop tools that can inform decision-making. While the Interactive Design Diagrams may provide a quick snapshot of design strategies, there are other tools that provide a more in-depth look at the evidence, allowing design teams to consider the information in context.

Prior to the graphic development of evidence portrayed in the Interactive Design Diagrams, the inpatient room content (medical-surgical, intensive care, maternity care) was developed through an extensive review of the literature and input from industry experts. Available on The Center's website as a PDF or Excel file, the resulting "checklist" and evaluation tools were tested for validity and reliability (Quan et al., 2017).

While tools for the deep diver typically start as an extensive narrative-heavy format based on the published research, the tools sometimes evolve into more user-friendly formats over time. The Safety Risk Assessment (SRA) toolkit (a proactive approach to identifying the latent conditions in the built environment) was developed through both the research and a committee of more than 100 industry expert volunteers over the course of 3 years (Taylor et al., 2012, 2014, 2015). Originally posted as PDF and Excel files (The Center for Health Design, 2015), the SRA was intended as a series of design considerations for decision making, but the format was difficult to use in a collaborative process due to the sheer volume of content.

A subsequent effort migrated the extensive content to an online platform (The Center for Health Design, 2017) that allows interdisciplinary teams to work together and enter project-specific information as the design evolves. The toolkit (see Figure 7) allows teams to consider multiple safety-related outcomes (i.e., infections control, medication safety, falls, patient handling, injury associated with mental and behavioral health, security).

https://doi.org/10.52202/067871-0008

Figure 7

Screenshots: The Center for Health Design's Safety Risk Assessment Toolkit

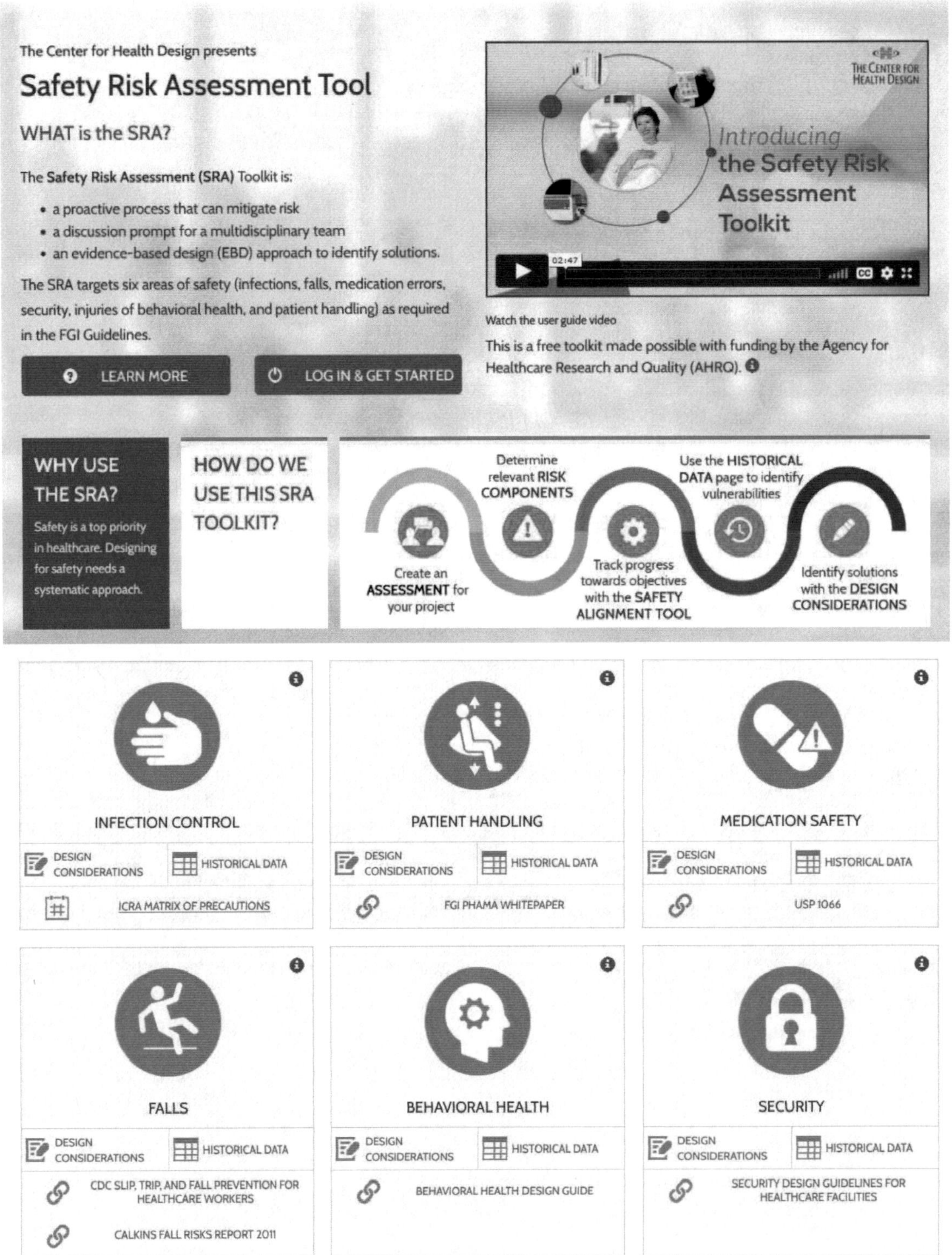

 https://doi.org/10.52202/067871-0008

As an interactive decision-making tool, the Safety Risk Assessment helps teams decide what is applicable to the project, what may pose the most risk, and what may be the highest priority for the organization. Design considerations are typically supported by cited research (Why? button) to answer questions about why the consideration matters. The online format provides space for teams to enter specific solutions, clarifications, or explanations, links to other shared documentation, and space to annotate additional alignments or potential tradeoffs (see Figure 8).

Figure 8

Screenshot: The Center for Health Design's Safety Risk Assessment - Design Consideration Example

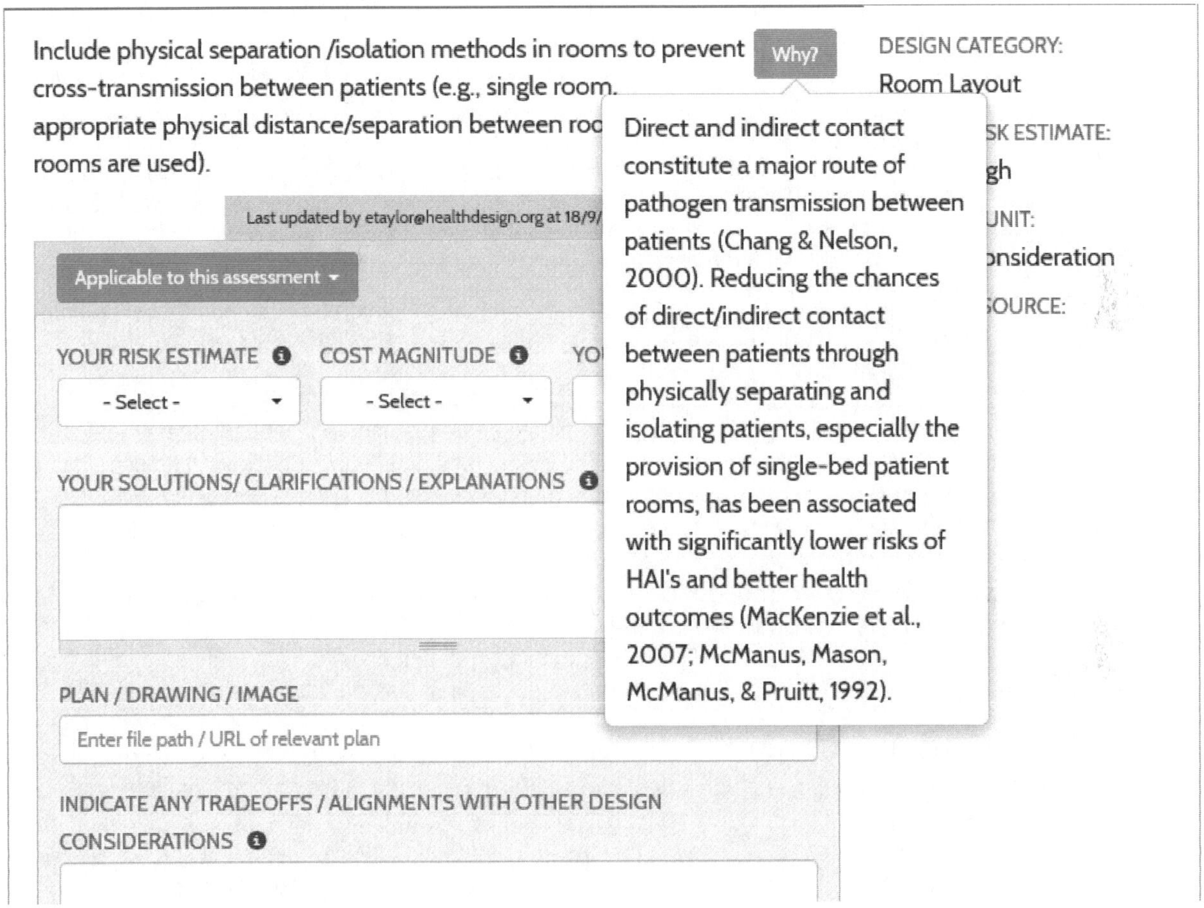

While the online SRA was tested extensively for usability, it still is best used with a deep-dive into all of the tool's features. [In the context of providing a time-saving and graphic format(tied to Opportunities 1 and 2) the user manual was created out of a series of short video tutorials.] The full functionality of the tool becomes easier to use and understand with each project.

Conclusion

Using research in design is a balancing act of time and resources to find, evaluate, and use the evidence. Teams need timely access to the appropriate information, which can be facilitated through the use of secondary research. Some secondary research outputs can be developed in an easy-to-understand overview for the surfer, whereas other secondary research outputs may warrant additional detail for the deep dive. While goals may drive the format, the development approach is often directly tied to available resources (e.g., staff time, funding). It must be recognized that just like any research, secondary research takes time to create (and use). The value is that the evidence does not always need to be (re) created as a custom enterprise for each project, and tools can continue to evolve with the knowledge base.

References

Arksey, H., & O'Malley, L. (2005). Scoping studies: Towards a methodological framework. *International Journal of Social Research Methodology, 8*(1), 19–32. https://doi.org/10.1080/1364557032000119616

Heinström, J. (2005). Fast surfing, broad scanning and deep diving: The influence of personality and study approach on students' information-seeking behavior. *Journal of Documentation, 61*(2), 228–247. https://doi.org/10.1108/00220410510585205

Huber, A. (2016a). Is seeing intriguing? Practitioner perceptions of research documents. *Journal of Interior Design, 41*(1), 13–32. https://doi.org/10.1111/joid.12067

Huber, A. (2016b). Research utilization in the design decision making process. *International Journal of Architectural Research: ArchNet-IJAR, 10*(1), 4–25. https://doi.org/10.26687/archnet-ijar.v10i1.824

Huber, A. (2018). Exploring interior designers' research utilization strategies and information-seeking behaviors. *Journal of Interior Design, 43*(2), 11–31. https://doi.org/10.1111/joid.12119

Internet Growth Statistics 1995 to 2022—The Global Village Online. (2022). https://www.internetworldstats.com/emarketing.htm

Nielsen, J. (2003, August 10). Information Pollution. Nielsen Norman Group. https://www.nngroup.com/articles/information-pollution/

Quan, X., Joseph, A., & Nanda, U. (2017). Developing evidence-based tools for designing and evaluating hospital inpatient rooms. *Journal of Interior Design, 42*(1), 19–38. https://doi.org/10.1111/joid.12091

Rabinovich, A., & Morton, T. A. (2012). Unquestioned answers or unanswered questions: Beliefs about science guide responses to uncertainty in climate change risk communication. *Risk Analysis, 32*(6), 992–1002. https://doi.org/10.1111/j.1539-6924.2012.01771.x

Rogers, E. M. (2003). *Diffusion of Innovations,* (5th edition). Free Press.

Syed, S. (2012). Recovery of gold from secondary sources—A review. *Hydrometallurgy*, 115–116, 30–51. https://doi.org/10.1016/j.hydromet.2011.12.012

Taylor, E., Card, A. J., & Piatkowski, M. (2018). Single-occupancy patient rooms: A systematic review of the literature since 2006. *HERD: Health Environments Research & Design Journal, 11*(1), 85–100. https://doi.org/10.1177/1937586718755110

Taylor, E., Hignett, S., & Joseph, A. (2014). The environment of safe care: Considering building design as one facet of safety. *Proceedings of the International Symposium of Human Factors and Ergonomics in Healthcare, 3*(1), 123–127. https://doi.org/10.1177/2327857914031020

Taylor, E., Joseph, A., & Quan, X. (2012). Designing for patient safety – Considering a patient safety risk assessment. In V. G. Duffy (Ed.), *Advances in human aspects of healthcare* (pp. 2620–2629). CRC Press. https://doi.org/10.1201/b12318

Taylor, E., Quan, X., & Joseph, A. (2015). Testing a tool to support safety in healthcare facility design. *Procedia Manufacturing, 3*, 136–143. https://doi.org/10.1016/j.promfg.2015.07.118

The Center for Health Design. (2015). Safety Risk Assessment Toolkit (PDF version) [Non-Profit 501(c)3 website]. CHD Insights & Solutions. https://www.healthdesign.org/insights-solutions/safety-risk-assessment-toolkit-pdf-version

The Center for Health Design. (2017). Online Safety Risk Assessment Toolkit | A Process to mitigate risk [CHD Tools]. *The Center for Health Design SRA*. https://www.healthdesign.org/sra

Total number of Websites—Internet Live Stats. (2022). https://www.internetlivestats.com/total-number-of-websites/

 https://doi.org/10.52202/067871-0008

This page intentionally left blank.

ABOUT EDRA

The Environmental Design Research Association (EDRA) is an international, interdisciplinary organization founded in 1968 by design professionals, social scientists, students, educators, and facility managers. EDRA came into being in the 1960s, during a period of social awareness and social unrest, but also new horizons and new visions. True to its origins, EDRA continues to believe in an agenda of social responsibility that has grown from a barely perceptible idea to a more fully-formed vision for healthier, more democratic human environments.

EDRA bridges theory, research, teaching, and practice to recognize and create more humane environments in collaboration with global stakeholders/communities, to promote cutting-edge research and practice that is essential to improve the quality of life for all people, to transcend disciplinary boundaries and industry sectors in a collaborative pursuit of environments that are responsive to diverse human needs, and to advocate for diverse voices in the pursuit of social justice. For more information on EDRA's programs and services, visit www.edra.org.

EDRA publications are just one of the ways to achieve EDRA's mission, and EDRA is grateful for Gensler's support of this compendium.

ABOUT Gensler

Gensler is a global architecture, design, and planning firm with 52 locations across Asia, Europe, Australia, the Middle East, and the Americas. Founded in 1965, the firm serves as trusted advisors to 3,500 active clients in virtually every industry. Guided by determined optimism, we believe the power of design can spark positive change and create a future that promotes equity, resilience, and wellbeing for everyone.

ACKNOWLEDGMENT

The world is just emerging from a global pandemic and is in the middle of several converging global crises - war, inequality, environmental degradation, and others. It has become imperative that environmental designers and researchers consider health - a state of complete physical, social, and mental well-being over time. EDRA has been trying to continue its legacy and address these challenges by providing an international platform to share knowledge and develop ideas. This publication is a timely and significant contribution to that endeavor.

I would like to thank Ann Sloan Devlin for her leadership in organizing this intensive, bringing eminent speakers, and producing this valuable publication. Her continuous efforts and dedication to EDRA are remarkable. I would also like to thank all the authors for participating in the EDRA53 conference and contributing to this publication. EDRA envisions publishing more of this kind in the future.

A special thanks goes to the EDRA53 conference organizer, Anjali Joseph, for plotting this idea to enrich EDRA's knowledge repository. I would also like to convey appreciation to her entire team for making the EDRA53 conference a success. I would also like to thank the students who supported the graphics of this report. A special thanks goes to Nick Watkins of Gensler for taking the initiative of sponsoring this publication.

Lastly, I would like to thank the EDRA Board members for their tremendous efforts to make this organization successful. A special thanks to Lindsay McCunn, the chair of the program committee, for working vigorously with the local conference team and making this conference a reality. I would also like to thank the executive director, Deni Ruggeri, for his dedication and support.

Sincerely,

Sharmin Kader, PhD

Chair of Board of Directors, EDRA